Working in Commercials

A Complete Sourcebook for Adult and Child Actors

Elaine Keller Beardsley

Focal Press

Boston   London

D1416546

Focal Press is an imprint of Butterworth-Heinemann.

Copyright © 1993 by Butterworth-Heinemann
ℛ A member of the Reed Elsevier group
All rights reserved.

Recognizing the importance of preserving what has been written, it is the policy of Butterworth-Heinemann to have the books it publishes printed on acid-free paper, and we exert our best efforts to that end.

Library of Congress Cataloging-in-Publication Data

Beardsley, Elaine Keller.
   Working in commercials : how to break in and stay in the
entertainment business / Elaine Keller Beardsley.
     p.   cm.
   ISBN 0-240-80160-1 (acid-free paper)
   1. Acting for television—Vocational guidance. 2. Television
advertising. I. Title.
PN1992.8.A3B38 1993
791.45'028'023—dc20   93-13433
                                          CIP

British Library Cataloguing-in-Publication Data
A catalogue record for this book is available from the British Library.

Butterworth-Heinemann
80 Montvale Avenue
Stoneham, MA 02180

10 9 8 7 6 5 4 3 2 1

Printed in the United States of America

□ □ □
□ □ □
□ □ □

I dedicate this book to my son, Nicholas, who had the good timing to be born on the same day we went into contract.

□ □ □
□ □ □
□ □ □

Think positively. If you think you are beaten, you are. If you think you dare not, you don't. Success begins with your own will—it is all in your state of mind. Life's battles are not always won by those who are stronger or faster. Sooner or later the person who wins is the person who thinks he can.

—Author unknown

Time makes a man find himself in his heart.

—Bhagavad Gita

Contents

Preface

The commercial business is very simple when you get right down to it. As you will hear time and again, show business, like any other business, requires focus, education, and method. While it is sometimes unpredictable and seemingly subject to flights of fancy, an actor can greatly increase his chances of success by knowing where he is headed and how to get there. While there are no ironclad rules, the information contained in this book is intended to provide enough in-depth and inside knowledge to enable the reader to make well informed and deliberate choices, instead of haphazardly throwing himself into the ring and making costly and time-consuming mistakes.

This book is a guide to the ever-changing and increasingly competitive business of television commercials. No other book written about the television commercial industry offers as much timely, in-depth information from so many sources. The novice, experienced, or studied actor will find the information it contains indispensable.

I have striven to offer a balanced and informed view of all facets of the commercial industry. Since I cannot help but be somewhat biased by my years as a television commercial agent, I have included several interviews with various leaders in the field to help present the fuller picture. These specialists readily shared their knowledge and expertise, and the highlights of these in-depth exchanges are presented.

Chapter 12, "Actor to Actor," is written by actors themselves. Those I interviewed have achieved varying levels of success and are at divergent points in their careers. While they may or may not be where they want to be vocationally, they are succcessful as individuals—which is no small accomplishment in this high-pressure, acutely competitive and ego-damaging industry. When asked the question "What have you done to pull yourself up when you felt that maybe you should try another career?" most asserted that they had never reached that point. When asked to give their impressions of the commercial business as a whole, and/or to impart any wisdom or advice they wished to share with other actors, they freely offered numerous tips to help others avoid the mistakes they had made. These perceptive and thought-provoking insights are reprinted virtually in their entirety.

Chapter 13 pointedly addresses in question-and-answer format issues that are more generally addressed in earlier chapters. Any industry professional who has given a lecture or taught a class knows that the same questions are asked over and over. The questions included in this chapter reflect actors' concerns.

While there are indeed no hard-and-fast rules, I believe the following are good guidelines for success in the entertainment field:

1. Have another interest to sustain you emotionally and financially. This will keep you anchored and prevent your becoming desperate.
2. Diversify. The more areas in which you excel, the more chances you have of making it. Skills and exposure in one area often overlap into another.
3. Study your craft, hone your talent. When opportunity happens, recognize it and be ready.
4. Don't take anything for granted. You may be flying high one minute and not having calls taken the next.
5. Keep your attitude in check and negative opinions to yourself. Memories are long in this business. You can't afford to alienate anyone needlessly, at least while you are still unknown.
6. Follow your heart and head, even if everyone is telling you something different. A combination of levelheadedness, knowledge, and intuition stand you in the best stead.
7. Do not be afraid of change or self-examination.
8. For every opinion you receive you will receive at least one equal and opposite opinion. Weigh everything and check several reliable sources before you proceed.
9. There are no guarantees, and sometimes no rhyme or reason as to why one person makes it and someone else doesn't. Another person's success or failure has nothing to do with you.
10. This business is in perpetual motion. Never stop working, studying, or improving your craft or yourself.

The odds of becoming a working actor today are not exactly inspiring, and those of becoming a star are overwhelming. Yet every day new and deserving souls break in, and some reach echelons they had only dared dream of. Many of the people who were just beginning to break into commercials when I started in the industry are now soaring to noteworthy heights in film, television, and theatre.

I wish to thank all these and other actors who serve as inspiration and proof that anything is possible when you put your mind to it.

Acknowledgments

I'd like to thank and acknowledge the following individuals for their special support and assistance in putting this book together:

NEAL ALTMAN	NANCY KREMER LOPEZ
HAY-DEE AMIANA	MARK LOCHER
JAMES BEARDSLEY	ELINOR LONDON
HELENA BROWN	NANCY LYON
RALPH BUCKLEY	BARBARA MASCARELLA
CARY CHEVAT	TAYLOR NICHOLS
VINCE CIRRINCIONE	JOHN MASSEY
BRIDGET COONERTY	JOHN McGUIRE
LEIGH CURRAN	KATHLEEN McNENNEY
CHRISTINA DEVRIES	FRAN MILLER
YVETTE EDELHART	DONNA MORAFF
DOUG ELY	KATIE MULLIGAN
LOUISE FOISEY	MARY GORDON MURRAY
DAVID FONTENO	CAROL NADELL
BETSY FRIDAY	TAYLOR NICHOLS
PATRICK GARNER	JONATHON PENNER
ROSE GARNER	JEFF PHILLIPS
TERRY GATENS	GREGORY POULOS
TRACEY GOLDBLUM	PATRICK QUINN
CAROL HANZEL	LETANYA RICHARDSON
GLENN JUSSEN	RANDY RUDY
JACK KELLER	JACK SANZONE
JOYCE KELLER	KAREN SCRIBNER
SCOTT KELLER	JOAN SEE
BARRY KIVEL	ANDREA SIMMONS
TOMMY KOENIG	DON SNELL

JOHN SPEREDAKOS   MAURA WALKER
LEXY SPETT   KATE WEIMAN
LINDA TESA OLKEN

In addition, a thank you to Mrs. Gershon and the research librarians at the West Islip Public Library

Thank you to Karen Speerstra, Sharon Falter, Louann Werksma, Jamie Temple, Maggie Dana, Mary Cervantes, and the rest of the folks at Focal Press for their kind and thorough help

A special thank you to the staff of the Drama Bookstore in New York City for their helpful information

A thank you and hello to my co-workers, friends, and mentors at Abrams Artists and Associates and the New York Casting Community

A thank you to all the actors who so graciously allowed their headshots or resumes to appear, whether or not they actually ended up in the book

An extra special thank you to Neal Altman and Fran Miller, whose valuable and unselfish input into this book cannot be measured

A special thank you to Tracey Goldblum, who always picked up the phone and was ready with a quick answer

A big thank you to Bridget Coonerty, without whose "Baby Wrangling" this book could not have been written

Sincere thanks to Professor Lauren Raiken at NYU for support and inspiration

Gratitude to my husband Jim, mother Joyce, father Jack, brother Scott, and children Alex and Nicky, for encouragement, moral support, babysitting, photocopying, mailing, and fixing computer ribbons above and beyond the call of duty

1

□ □ □
□ □ □
□ □ □

The Commercial Industry

Many actors go through drama school or intensive acting technique training programs bent on careers as legit actors, only to be confronted with a brutal and unappealing reality: Making a living as an actor can take years, and in some cases a lifetime, to accomplish.

Commercials are not presented as artistic alternatives, and training in this area is usually not included in the curriculum of acting or theatrical schools. In fact, career guidance is rarely included in the dramatic course of study, which leaves a serious gap in the education of someone who intends to make a living as an actor. The fact of the matter is, if you don't reach quick success as a performer, you may be facing years of financial and emotional struggle in an unsatisfying, and probably unrelated, job pursuit. Since "survival" jobs are time consuming and not usually lucrative, they serve to infuse the actor with a sense of helplessness and desperation—the two surest career killers around.

While commercials are just that, *commercials*, and entail using your skills to sell a product or service, they offer a viable and attractive route to success for even the actor with the highest artistic ideals. They no longer stigmatize, ruin careers with overexposure, nor lessen an actor's caché or esteem within the industry. Rather, they present the opportunity for financial and emotional stability, leaving the performer free to study and to pursue less financially rewarding acting endeavors. Income from commercials can provide actors the flexibility to pick and choose in their careers, enabling them to pass up less-than-attractive offers for more strategic and satisfying moves.

Commercials also allow actors to hone yet another aspect of their craft. Through volume of auditions and brevity of material, they require performers to make quick and decisive choices, a skill that serves them well in all facets of their professional lives. Commercials foster contacts. They keep an actor's talent fresh and energy up between *legit**, or film, theatre, and television auditions.

---

*The term *legit*, as it is used in the entertainment industry, refers to film, television, and theatre, and the actors who work therein. It's use is not intended to relegate commercials to an inferior position, but rather to delineate the commercial and non-commercial areas of the business.

Since there is so much more volume and turnover in this area of the business, commercial agents, casting directors, and directors are far more receptive to untried talent than their legit counterparts. There are more jobs and opportunities for work in commercials. Industry professionals who work in commercials often cross over to the other areas—casting, agenting, or directing for film, television or theatre, either simultaneously or at some other point in their career. The actor in their good stead will likely be called upon again and again for whatever project they may be working on.

Commercials offer exposure. Many actors are spotted this way and considered for work on other projects. Steven Spielberg called in one of our actresses after seeing her on a yogurt commercial. After a lead in one of his films, she went on to a contract role in prime-time television.

Yet commercials are not solely a means to an end. They can be gratifying and rewarding in and of themselves. Many actors phase out other areas of their careers after finding success and satisfaction in this corner of the business.

Let's face it, commercials have enmeshed themselves in American, and international culture. They are part of our everyday lives, influencing us in ways of which we may be totally unaware. While in the past they may have been branded as boring or uninspiring, today's commercials are reaching new heights of ingenuity, creativity, and originality. In many cases, commercials are making inroads that influence the rest of the entertainment industry. Commercials have evolved from the straightforward, uninteresting hard-sell of the days of live television, to high levels of art, entertainment, and meaning. While a commercial's main goal is indeed to sell products, an actor with vision, drive, and integrity can work in and around its product-based nature and come out with creative standards intact.

The commercial industry can be available to anyone who is personable, aggressive, confident, and knowledgeable, who remains true to self and maintains a sense of perspective. This arena is always open to good actors and people with new and exciting faces. If you hear the words "We don't need your type," you are doing something wrong. But before I tell you how to do it right, let's look at how the industry works, who the players are, and what the audition process is.

The Commercial Process

The client company engages an advertising agency to plan and implement a single commercial or campaign. The creative team, which consists of the agency, creative director, copywriter, art director, and possibly others, formulates a concept, scripts it, and visualizes it on a *storyboard*. If the agency team needs help to convince the client an idea will work, they may produce a *demo*, a less than full-scale version of the final commercial. In some cases, this version will be of high enough quality to actually show on air.

Occasionally the agency and client will be so taken with the performer hired for a demo that they will use the same talent for the commercial without auditioning anyone else. In any case, the performer who appears on the demo stands a good chance of being considered for the actual commercial.

After the client and agency agree on a plan of action, the production team takes over. The producer selects a director, film house, and casting director. Sometimes the ad agency will cast in-house through its own casting departments, but as agencies merge and budgets are slashed these departments are in the minority. Often an independent casting director is employed, particularly if the ad agency is located outside the three major commercial areas—New York, Los Angeles, and Chicago.

The casting director serves as liaison between the client and the talent. After interpreting the casting needs of the commercial, she contacts agents, or, in some cases, the talent directly, using her own knowledge of the industry and other pertinent information. This information includes the amount of money available, the quality of the commercial, the area where the commercial will be shown, the cross-section of the population to be attracted, and the particular likes and dislikes of the director. Most casting directors have a handful of agents they feel comfortable working with, who respond to the casting director's needs honestly and appropriately. It is a working relationship like any other: The agent needs the casting director to call with jobs, and the casting director needs the agent to supply her with the best talent available. The agent, whose job it is to locate work for clients, prioritize, and negotiate for them, makes a number of suggestions.

The casting director sets up a session, which could last a few hours or a few days, depending on the number of characters involved, amount of copy, and overall difficulty level. A commercial may be difficult to cast for any combination of reasons. It may be written poorly or not offer compensation above union scale. The decision makers may be finicky, they may not know what they are doing or exactly what they want, or they may not be able to articulate their needs. The concept may not work and thus be extremely difficult to cast, the director may have very specific taste that is hard to satisfy with existing talent, or it may be off season when all the actors are out of town. It could be a commercial that seasoned actors find undesirable: Many actors object to doing television commercials or print ads for hemorrhoid medication, feminine protection, furs, alcohol, cigarettes and so on. It might be in a high *conflict* area. While all commercials have conflicts (for instance if an actor appears on a deodorant commercial, she may not appear on a competing product running in the same area) one may be in a higher conflict area than another. The types of roles the actor primarily auditions for dictate her areas of highest conflict. If you are a young, good-looking guy, beer would be a high conflict area for you. Laundry detergent for homemaker types, soda and candy bars for teenagers are other examples. In the case of these and other high conflict areas, the agent for the more experienced actor will want to insure that it is running well (in one or more major market) or at least secure some extra money for her client. If any of these factors come into play it could take extra time to cast the commercial.

The casting director picks and schedules actors for the audition session, tapes the audition, and sends the audition tape to the advertising agency for review. Often, quite a few decision makers are involved: the client(s), director, producer(s), creative director(s), and possibly others. The choices are narrowed down, ideally to two or three performers per role, but sometimes many more. *Callbacks* are then scheduled and *first refusals* given out. A callback will be attended by some or all of these decision makers. A first refusal is when an actor holds his time open to insure that he is available on the projected shooting date.

The commercial is then shot, edited, and distributed to the stations. Often it is on the air in a matter of weeks.

What's In

For years the commercial market relied heavily on the Proctor-and-Gamble look, commonly referred to as *P&G*. This is the all-American, white-bread, cookie cutter image exemplified by Donna Reed and Meredith Baxter-Birney. Although certain companies still utilize this type to sell their products, there has emerged a welcome trend towards the individual. Actors who in the past have been branded as *ethnic* or *interesting* types are currently in vogue. The person who heretofore was not a commercial "type" can now find work in this business. Progressive commercial directors such as Steve Horn and Joe Pitka are credited with spearheading this movement towards the individual.

Although the P&G type still commands much work, there is a growing market for the "offbeat" or ethnic type. Furthermore, there is a widely growing Hispanic market on both network and Spanish language television. Actors who *look* Hispanic would do well to learn to speak Spanish.

Stand-up comedians and actors trained in comedy and improvisation are well received. Although the audience may not realize it, many of the actors cast in even slightly humorous commercials are comedians. Whether it shows up in the final version of the commercial or not, humor in general goes a long way in furthering an actor's career. Everybody likes to laugh, and casting directors are no exception. Given the choice, they are more apt to include someone in their casting session who will brighten their day. The decision-makers viewing the audition tape are likely to be drawn toward the legitimately funny and inventive individual and may alter their view of casting the commercial. This does not suggest you force humor by making time-consuming or inappropriate jokes. If you have comic ability, develop it. If opportunity allows, incorporate humor into your work.

Models who can act are in demand. If you are a print model looking to break into commercials, widen your market by taking legitimate acting classes. The ability to speak well and emote in an easy and non-stilted way will dramatically improve your salability.

The commercial market can be one of mimicry. Whatever is hot on television or film is used to sell products. At one time, Tom Hanks, Kirk Cameron, "Thirtysomething" types, Bruce Willis, and "LA Law" types were in. In fact, the popularity of Bruce Willis and Corbin Bernsen was largely responsible for the advent of the leading man with a receding hairline. Previously Ted Danson, Shelley Long, Michael J. Fox, and *Big Chill* types were hot. Mariette Hartley, James Garner, Goldie Hawn, and Teri Garr are types that have remained in demand for years.

Keep an eye on the box office and TV ratings. *Variety*, *Ad Age*, and *Advertising Week* are also good indicators. Do your homework and soon you will be able to anticipate the trends. Keep in mind, though, that the people casting commercials are not looking for an exact replica of the star. A prototype merely indicates an appeal, manner, or general look. For instance, a Goldie Hawn type usually means attractive, slightly ditsy, but intelligent. Often the actor cast in the commercial bears little or no physical resemblance to the prototype.

What It Takes

The commercial industry is a business like any other. The actor, having reached a level of confidence, ability, and self-discovery, needs to market himself. This first involves finding a niche and then targeting that area. Picture chiseling a hole in a wall and then widening it bit by bit.

As agents in the business like to say, "There is no such thing as a non-working actor." Instead of waiting for something to happen, the actor must *make* things happen. This takes a lot of legwork, energy, and, unfortunately, money. It can take as long as one to two years before you book your first job, and booking your first job offers no assurance that you will work consistently. So, not only will you need fortitude, self-confidence, and desire to succeed, but the desire has to be all-consuming to get you through the rough periods.

In addition, an alternative source of income is absolutely essential not only for basic expenditures such as food and rent, but also to cover professional expenses. A good photograph and resume can easily cost $1000. Add to this the cost of a wardrobe necessary for selling yourself as your type, hairstylist, postage, telephone bills, trade papers, and transportation.

Characteristics I've found to be desirable for success in commercials include

talent
specific looks
charm
charisma
confidence
ability to handle rejection
gregariousness
salesmanship
ingenuity
comedic ability
overwhelming desire to succeed
strong sense of self
ability to take criticism
willingness to take constructive action
boundless energy
dauntlessness
desire to improve
willingness to sacrifice
candor
ease with people
ability to deal with the fickleness and general unfairness of this business

If few of these qualities apply to you, think twice about the commercial industry. Perhaps it is not the right time or the right business for you. Otherwise, read on.

___

Interview with Linda Tesa Olken, Commercial Producer
Former Vice President/ Director of Television and Radio Production
Della Femina McNamee, Inc.
350 Hudson Street
New York, NY 10014

BEARDSLEY: As someone who has worked directly with commercial production for more than 13 years, can you tell me how the advertising world has changed?

OLKEN: The big change has probably been the emphasis on money and the economy. Clients are cutting back on advertising right now.

BEARDSLEY: What are they advertising with? Or are they not advertising at all?

OLKEN: Radio production has boomed because production costs are much cheaper.

BEARDSLEY: Are they reinstating old commercials?

OLKEN: Many clients are using commercials produced in the past. Several years ago I produced some original music for a major bank. They are still using it today. I don't know if they just love the music or if it's a statement about the economy, but we seem to be seeing commercials running for a lot longer than in the past.

This has affected the acting community as well. There are fewer and fewer jobs for actors. Many actors who in the recent past required double and triple scale are glad to have any work at all.

BEARDSLEY: Are there any areas that are better than others?

OLKEN: The trend seems to be away from comedy. Now we are seeing a number of commercials that use family-oriented, slice-of-life situations. There seems to be a back-to-basics approach to advertising.

BEARDSLEY: Are they going back to straightforward types?

OLKEN: The trend has been to cast real people for real-life situations. In the beauty/fashion category, the trend has been to cast the 1960s look. The 1960s "grunge" look has hit advertisers.

BEARDSLEY: How do you choose a director for the spot?

OLKEN: Usually we select three or more production houses to bid. The process begins by screening reels. The most important aspect is to match the director to the type of spot. Some directors are great for comedy, others are great for tabletop, and so on. We try to keep an eye out for up-and-coming directors who may have a fresh approach to the material.

BEARDSLEY: Can you identify some of the newer directors?

OLKEN: David Wild comes to mind. He is very funny and very natural.

BEARDSLEY: What's the length of time between a commercial's conception and getting it on the air?

OLKEN: Of course it varies according to the complexity of the spot, but the general rule is eight weeks. This process includes: screening directors' reels, bidding, estimating, client approval, preproduction, postproduction, editing, and, finally, shipping to stations for airing.

BEARDSLEY: How hard is it to get the client to approve the commercial?

OLKEN: Making a commercial is such a complex and expensive undertaking that clients, understandably, want a high comfort zone before they commit. Producers try to make the client a partner in the endeavor and include them in the decision-making process. For example, you would not present a storyboard to a client using a celebrity spokesperson without checking if that person is available or without checking the approximate fee that person would charge. The more homework you do prior to presenting goes a long way toward getting the client to feel comfortable with and committing to the project.

BEARDSLEY: In terms of talent, how close to the original idea is the final product?

OLKEN: We try to come as close as we can, but in the casting process, sometimes, things change. *We* have a preconceived idea of what we want, but when we see someone who gives a great performance in an unexpected way, we may decide to use that person, thereby differing from the original idea. At the outset, we set some specific criteria for the casting director to follow and she usually comes pretty close to our original conception. We usually present one or two choices to the client.

BEARDSLEY: Do you ever present more than that?

OLKEN: Rarely. We would only present additional talent if a problem were to arise.

BEARDSLEY: In the end do you generally agree?

OLKEN: Yes. The creative people, the director, and the client work as a team.

BEARDSLEY: How much input does the casting director have?

OLKEN: There are several excellent casting directors I have worked with over the years. People such as Donna DeSeta, Karen Kayser, and Joy Weber. These people are true artists. They understand what you are looking for and are meticulous about finding the right kinds of people. They don't waste your time with cattle calls in the hope that you'll find someone you like. They really work with you to add value to the spot. They make the process much more enjoyable.

BEARDSLEY: Is there anything that you would like to say to actors?

OLKEN: Only that they should study the material thoroughly, listen to the casting director, and try to be as natural as possible. They should follow their hearts in reading the scripts.

BEARDSLEY: Can you describe what it is about a reading that makes it right?

OLKEN: It really just happens. It is impossible to describe what happens in the room when an actor hits it just right. Smiles just start to light up on everyone's face. It makes it all worthwhile.

Linda Tesa Olken served as the on-line producer for Pan Am, Raytheon Corp., Chemical Bank, Beck's Beer, National Institute on Drug Abuse, and other accounts. Her work on WCBS-TV garnered her an Emmy for Outstanding Promotional Announcement.

2

□ □ □
□ □ □
□ □ □

Perfecting the Product

Most people in the entertainment industry generally agree that the key to gaining success as an actor is having a certain something, a *je ne sais quoie*. Some call it star quality, charisma, or magnetism. Others refer to it simply as a sparkling personality. It is a certain essence that defies explanation and causes heads to turn when such a person walks by. It is communicated through bearing, voice, expression, or gestures. It can be a regal bearing, an attitude of aloofness, or a carefree quality. No one knows exactly what it is or where it will turn up next, but, when it does, that person, talented or not, will be in demand. Some will become stars.

This is not to say that if you don't possess that ineffable something you won't work. You merely have to work harder on the things you can control. The first of these is *attitude*.

Attitude

Turn-offs

Possibly the first thing that directors, agents, and casting directors will pick up on is your attitude. It is transmitted through your phone calls, your cover letter, sometimes your photograph, and always through your presence. The actor with a chip on his shoulder is a tremendous turn-off. He sends out a signal that says "No one will hire me so there must be something wrong with them." He has convinced himself that it's another person's job to discover him, not his responsibility to get discovered.

Another turn-off is the sometimes subtle, but whiny "I can't understand why no one will hire me" presentation. This actor has convinced herself that she can't find work because of factors beyond her control. Whether this is the case or not, an actor who wants work will accept responsibility for her actions and stop complaining about things not going her way.

Turn-off #3 is the presumptious "Here I am, now you can stop taking your calls and give me your undivided attention" egomania. This person is put out if asked to wait, produce a picture, read copy, deal with assistants, or do any of the other lowly things that actors fortunate to be free of such an attitude are willing to do. Although many celebrities are afflicted with this particular attitude disorder, it is also prevalent among as-yet-undiscovered talent.

Some other attitude turn-offs:

- The depressing "I'll never work again."
- The befuddled "I can't understand why I can't get hired, or why so-and-so always gets the job, or where I fit into this business."
- The artificial "I'm so happy I could just burst."
- The antagonistic "This business, or that person, has done me wrong."
- The arrogant "I know everybody and everything about this business."
- Write in your own_____

Although these descriptions may seem to border on the caricature, they are in fact quite common and can be found in varying degrees among actors seeking work. However subtle, they are swiftly perceived by the very person you are trying to impress. While you may be occasionally fortunate enough to find sympathy or a willing ear with some of these dispositions, you will probably not find a job.

Turn-ons

Although your attitude may be justified or difficult to keep in check, you must find a way to keep yourself psyched up and focused on reality. If you are not coming across as though you are on top of the world, secure in your talent, and thrilled to be part of this exciting business, work on it. Attitudes are infectious. If you are not sure of yourself or your talent, you can't expect others to be. There is nothing more uninspiring than an actor in despair. On the other hand, many an actor has overcome the odds with a winning optimism and charm.

Realize that the business is unfair. Those who are on top one day may be down the next, for no specific reason. The people making the decisions in this business do make mistakes.

The following story is true and similar situations happen more often than we like to admit:

About four or five years ago an actor came into our agency who was singularly unimpressive. He was low-keyed and had no credits to speak of. To make matters worse, he was not that attractive and a tad slovenly. The agents vowed revenge on the manager who sent him. Today, this young man is one of the hottest teen idols to ever hit television, leaving one hit show to headline a second.

Apparently this young man had a certain appeal or talent that we missed.

Realize that if you genuinely do have talent and persevere, chances are you will be found eventually.

Find a way to psyche yourself up before interviews, auditions, and phone calls. An upbeat positive attitude is the only kind that works. It has a way of permeating the air around you, and those who come in contact with you can't help but be affected by it. Be wary of overdoing and coming across artificial or insincere. Don't get carried away. Be genuine.

Heat

All actors, even the employed ones, go through periods of depression. It's an insidious thing, infecting auditions, your work, and your dealings with people. It can and does cause a further downslide of careers. Once in it, it is extremely hard to break out of. The greatest proof of your acting skill is your ability to leave despondency at the door. It is essential you master this skill because actors who are flying high seem to influence everyone they meet with their positive attitude. The high they get from giving a good audition or obtaining a job permeates their space and helps them secure the next success.

Attitude is one of the key ingredients of *heat*. For those not familiar with the term, it generally describes actors whose careers are accelerating at warp speed. This elusive condition is desired by everybody but obtained by few. It can last a moment, or a decade. It rarely lasts an entire career, but when you have it, everybody wants you. Unfortunately, it can end at any moment with little warning and no explanation.

Use heat to your advantage if you should be so lucky to experience it, but remember to keep a level head and your wits about you. Try not to let this or the low periods affect you or your work. Remember that the fickleness of this business has little to do with you.

Appearance

The next thing to work on, and of particular importance in the commercial arena, is your appearance or look. Some people believe that this is *all* you need, but in reality it is only one part of the whole package that is you. Your look might get you a job, but it will not make your career. If you haven't perfected your look and figured out your place in the market, chances are no one else will, either.

Firstly, and most importantly, do whatever is necessary to give yourself confidence. If five extra pounds keep you from getting up in the morning, lose them. On the other hand, certain character actors benefit from some extra weight and find that they don't work as much when they are too thin. This all comes from knowing your place in the market.

Secondly, do a complete rehaul *before* you take pictures and embark on your show business career. It is best not to make any radical changes without some professional guidance, however. Get your teeth fixed and your hair styled and colored, if necessary. Get into shape, if it will help. See a make-up expert if your make-up is not doing what it should for you. There are companies who specialize in image consulting, helping to choose the hairstyle, clothes and colors best suited to your personality. They are relatively expensive, but for actors having a hard time putting themselves together they are well worth the money.

Finding Your Type

Those actors who arrive in New York or Los Angeles believing they are the boy or girl next door are usually rudely awakened. Knowledge of your specific type and where you fit into the market generally comes only after a long process of trial and error and self discovery. For success in commercials, you must know yourself and your competition well enough to know how you are alike and how you are different. The people you encounter will try to determine where you fit in the market so that they can peg you, but the thing that will get you the job and give you longevity in the business is your uniqueness. Actors are often disturbed by the tendency of the market to pigeonhole performers by type, yet they allow themselves to be victimized by this. It is up to the actor to bring his versatility and talent to the forefront after generating initial interest as a perfect type. As more and more people get to know you and your talent, your salability will increase. In all fairness to the commercial industry, lines have been blurred as an increasing array of people are perceived as middle-American.

In the beginning, at least, an actor must determine his place in the commercial market. He does this by watching television and identifying the different types seen there. He talks to people in the business (not his relatives) and gets their opinions. He networks, talks to commercial directors, casting directors, and agents, and gets their opinions. Agents and casting assistants are often a good source of information. In short, talk to as many people in the know as possible. You are sure to get some overlapping opinions. Some types will be obvious, such as the "handsome spokesperson." For others it will take a great deal of experimentation. What follows is a broad overview of commercial types and related terms:

Spokesperson: Usually middle-American looking, can be ethnic, usually attractive and on the conservative side, always well-spoken with no accent, a voice with a clear, crisp quality. Spokespeople are professional in look and attitude and always able to handle copy well.

Grandpa, Grandma, Mom, Dad and the kids: Warm, character to attractive, mid-American to ethnic

Teenager. Cute or cool, can be trendy, usually spunky, usually attractive

Blue Collar. The average working person from construction worker to waitress, can be real or character, age 20–60

Career Person. Anything from straight to a character actor, but with a businesslike air, age 20–60

Beauty/Glamour/Model. Male or female, all ages, must be extremely attractive. Usually only professional models are used on these jobs unless there is a lot of copy.

Girl or Guy Next Door. Real looking, teens to 20s, can be straight looking or on the character side

Comedian/Comedienne. Used more and more in the traditionally straight roles such as that of young parents and career people

Real. As opposed to extremely beautiful or glamour. The kind of person you would see walking down the street

Ethnic. Anybody not whitebread. This includes actors who appear to be Italian, Mulatto, Jewish, and so on.

Whitebread: What in the past has been known as P&G (Procter and Gamble), very white, without any hint of ethnicity

Off-Beat: Not your typically straightfoward type. Has an "edge"

Having an Edge: Something that distinguishes this person from the crowd, someone having no "soft corners," can be straight or character

Off-center: Another term for off-beat, depending on who is using the term

Left of Center: Straight but with an edge

Straight: No hidden flaws, no surprises. Mr. or Mrs. Perfect. All soft corners

Character: Can be heavy, ethnic, balding, skinny, nerdy, or basically anything that keeps this person from being classified as straight

Upscale: Upper class looking, usually more attractive than real, can be character

Street: Your typical Levi's wearer as perceived by the media: hip, trendy, cutting edge. Often able to do neat things such as skateboard or play the sax. Often ethnic, teenaged and older

Rugged/Outdoorsy: Lumberjacks, mountain climbers, Ivory girls, Marlboro men, basically anyone who looks like they belong in a pair of hiking boots or in a saddle

Wardrobe

After you've determined where you fit in, do everything in your power to perfect that look. Develop a wardrobe, starting with a few basics, then expanding it bit by bit. Generally speaking, stay away from black, white, busy patterns, and with the possible exception of some teenagers, trendy or radical clothing or hairstyles. Here are a few essentials for each character type:

Spokesmen: A suit and tie are required, hair should be styled conservatively. Men who wish to keep their hair long so that they can also fit into the more trendy parts or do classical theatre, can try pulling their hair into a ponytail and tucking it into their shirt collars. This may work for a time, but somewhere down the line you will be probably be forced to make a choice. Think Bryant Gumble or Alan Thicke.

Spokeswomen: You will need a suit or a jacket and skirt, a blouse with a bow and a button down shirt for corporate products. You should have a couple of coordinating scarves to soften or dress up the look. Jewelry should be simple and subtle, hair should be conventional and above the shoulders. Putting your hair up into a bun may work for some spots, but for the most part this tends to make the wearer look frumpy or dated. Think Diane Sawyer or Phylicia Rashad.

Parents and Grandparents: A plaid or casual shirt for a downscale look; a nice shirt and slacks or a skirt for a casual look; jacket and tie, or dressy blouse or dress, for a more upscale look. You should be prepared for all three.

Teenager: You should have at least one trendy outfit and one nice casual ensemble, such as a shirt, sweater, and slacks or skirt.

Young Mom: Something soft and maternal, such as an angora sweater or a blouse with a gentle bow, projects this image well.

Young Dad: Shirt, sweater, jacket, and slacks

Blue Collar: Men will need a denim shirt. Women: denim, plaid, or plain shirt will do. Simply suggesting a uniform is enough, wearing a full waitress uniform or garage overalls is not necessary on an audition.

Career People: Men will need a jacket or suit and tie, women a jacket, blouse and skirt. The recommendations for spokespeople apply here as well.

Glamour/Beauty/Models: As in the other categories, simple is best, but more trendy or contemporary clothes work well here. There is more leeway with the hair as well, as both men and women can wear it either long or short. Men beware: Long hair will limit your being accepted for other types of roles.

Girl/Guy Next Door: Plaid shirts for both sexes go over well here, as do most types of conservative casual clothing.

In general, everyone should possess an *upscale* and a *downscale* outfit (dressy and casual). You can refine your wardrobe as you discover what type of auditions you are going out on. Remember, the average American consumer must be able to identify with you.

Hair

As mentioned before, it is better to have a conservative style and cut than to limit the types of calls you can go out on with a trendy or radical style. If you are a teen, model, or in your early-to-mid 20s with a contemporary flair, you could try a versatile style that goes either way. People who color their hair beware of roots; they are greatly magnified on camera.

Men

Men with receding hairlines should invest in a *good* hairpiece or wear none at all. The balding look is in, so a toupee is not essential; but having one on hand will broaden the spectrum of roles you can audition for. Hair plugs are expensive, permanent, and time consuming, and you cannot work or audition if you are in the process of having them done. Hair weaves are natural looking only if done properly, which is rare. Your best bet is probably to go natural, unless you tend to audition mainly for straight spokesman roles. In this case, a good toupee is certainly worth the investment. Have photos taken with and without the hairpiece.

As a rule, beards and moustaches are not in, but if you look good in one and can grow it quickly then have pictures taken with and without. Some ethnic men do very well with facial hair. For men with a specific look, such as "sea captain" or "Marlboro man," beards or moustaches can work well. Make sure it compliments your look and does not overwhelm your face. Realize that it may limit the amount of calls you can go out on.

Gray hair often works well on men, bestowing authority or a distinguished air. Don't be too quick to color. Many a younger man has started to work because he went gray.

Women

Older women do well by lightening their hair as they start to gray, but be careful of doing it for too long. Gray hair may actually compliment your look and help you to get hired. Many younger women who have gone gray prematurely have found that they are winning jobs over older women. It all depends on your look and age range.

Make sure your hair color and style compliment your overall look. Older women who insist on keeping their hair long limit their salability. If you have a dark complexion and dye your hair blond because you are trying to fit into more roles, you may find that you can't get hired at all.

Men *and* women: know your type and where you fit into the market. Don't cling to old and comfortable styles that may not be working for you. Be flexible, reasonable, and open to suggestion, or you may never fully perfect your look and realize your full commercial potential.

Commerical Classes

Commercial technique differs from acting in film or theatre. Everyone just beginning in this medium, including professional actors, can benefit from commercial instruction. Classes in general are a great way to keep the juices flowing, make contacts, network, and brush up on technique. They are absolutely essential for those who have never acted. Seasoned actors need not take a whole course, but may avail themselves of a one-on-one situation, or a couple of brush-up classes. Don't assume that because you are proficient in other areas of the business that you will be able to perform in commercials. You are playing to an entirely different audience. You are selling a product, not yourself. Some of the most successful commercial actors are not necessarily the best actors, and terrific actors are not automatically skilled at delivering commercial copy. However, the best and most sought-after commercial performers are talented and trained actors. Think of this as honing one more aspect of your craft. The more areas you perfect, the more well-rounded and marketable you will become.

Ask casting directors, agents, assistants, and other actors which classes are the best, then audit as many as you can. Some don't allow auditing, but that doesn't mean you should scratch them from your list. Generally speaking, classes that screen prospective students are better than those that don't. Some don't accept students who have no acting experience. Check class size, the background and experience of the instructor, the caliber of students taking the course, and general perception within the industry. Most good instructors and classes are well known by their peers. Some of the most popular and expensive schools are considered a waste of time and money by industry people. *Check around.* Do not go to the school that does the most advertizing or makes bold promises.

Some commercial classes are taught by working casting directors and some bring agents and casting directors in at the end of the course to see your work. This is terrific if your work is on a professional level, or in some instances if you have a wonderful look. But for the actor whose work is not ready to be seen, this can leave a damaging first impression that will be next to impossible to overcome. Many agents regularly do pick up talent from commercial classes. You might ask potential instructors how many of their students have gone on to become signed clients as a direct result of being seen in their class. Ask if you can contact former students for information.

A good class will give you plenty of time in front of the camera, will allow you to experiment with different types of copy, and will allow you to make your own acting choices within the boundaries of the medium. Class size should be relatively small. A videotape of your in-class work is unnecessary, unless you want it strictly for your own use.

Commercial class ripoffs abound. Check with the Better Business Bureau and the State Attorney's office before giving anyone any money. Call your local union office (see Appendix) for names of franchised agencies in your area, then call and ask for their input. They are often happy to help.

Pictures

Every actor knows how important it is to have a headshot, but a look at any casting director's or agent's mail will show that few know anything about getting a good one. Having the right picture is invaluable. Sending out a bad one will do you no good and may even hurt you. But like anything worth having, getting a terrific headshot takes time and effort.

Unless you are well known, have been picked up from a show or class or personally recommended, your headshot is your entree into an agent's office. If it's bad or out of date, it may end up in the garbage. If it's misleading and presents you as something you're not, when you walk in the door the person you are meeting with will, at best, be disappointed. It may go into a file to be pulled later for extra work or some other specific reason, at which point you may be completely wrong for the job—wasting time and creating bad feelings toward yourself. In short, an accurate, engaging shot is vital and worth every penny.

The Photographer

Paying a lot of money or going to the hottest photographer in town will not insure a good headshot. These people may have long waiting lists, possibly as much as two weeks for a meeting and six months for the actual sitting. Going to someone unknown does not mean you will get a bad shot, but it does require more caution.

In any event, ask around. Most experienced actors can rattle off the names of the better known photographers without batting an eyelash. Many talent agencies have referral lists. You need not be a client to call and ask for some suggestions. Many agents actively seek out new and upcoming photographers to send talent to as the established ones become more popular and expensive. This is not to undermine the expertise and know-how of the tried and true, however. They have worked long and hard for their reputation and may well be worth every penny. Your best bet is to get referrals from an agent or casting director who knows you and the look you want to project. Many photographers have a special ability in a particular area, such as beauty or children, but are not great at character or real shots. Check around.

Next, set up meetings with at least two photographers, three or four if this is your first time. Get their price lists and exactly what is included in writing. A headshot can cost anywhere between $200 and $900, not including reproduction costs. Many times as photographers gain popularity, their prices go up. If you get a strong recommendation for someone new and upcoming, look into it.

Make sure to look at the photographer's headshot books. Look for vibrant shots, where the personality seems to leap off the page. Look for freshness and different approaches. If all the subjects look alike, if they are all posed the same way, or if the lighting is always the same, chances are this photographer is inexperienced, uninspired, or stale. Check for the type of look you want to portray. Ask the photographer what types of shots she likes to take, what outfits she recommends, what backgrounds she uses. Some photographers like to shoot outdoors or use a hand-held camera, but very few can do this effectively.

If you are unable to establish rapport or feel uncomfortable with the photographer, then pass her up. It is essential that you feel at ease, or the tension will come through in the photographs.

Beware of the photographer who makes all her subjects look overly glamorous or too attractive. Many photographers have built an extensive clientele for this very reason, but these shots tend to gloss over the real person. While all actors should look their best on film, not looking like your photo will do you no good. It may antagonize the very people you are trying to impress.

Practiced photographers have their own distinctive style, and experienced agents will be able to steer you to the right one. Remember, though, the choice is yours.

What You'll Need

Everyone should aim for one smiling commercial shot to begin with. In New York, agents and casting directors like a headshot. In L.A. and for commercial print on both coasts, a *composite* is necessary. Gaining in popularity is the three-quarter head and body shot which includes the head and most of the body (see Fig. 2.11 and 2.18). The full or three-quarter shot is popular for soaps and is generally accepted commercially. If you want this type of photo, ask your photographer to see samples of her full and three-quarter shots.

Start by getting at least one good commercial shot and build up from there. If you take on too much to begin with, you might end up with nothing at all.

Getting the Perfect Headshot

A good headshot, be it for commercials or legit work, will convey

naturalness
personality that leaps off the page
eyes that are alive and vibrant
a smile that is engaging and real
the real you
energy

Furthermore,

Your teeth should be showing in at least one chosen shot.
Your whole face should be in focus.
You should be facing the camera.

Watch out for

a forced smile
smiling eyes with unsmiling mouth
smiling mouth with unsmiling eyes
squinty eyes
distracting clothing, jewelry, or background
stray hairs

Stay away from

arms or other body parts in the picture, unless it is a full or three-quarter shot
accessories, unless it's a character shot

Outfits worn for the camera should not distract from the face. A simple
neckline is best, depending on the look. Here are some suggestions:

collegiate young men and women: a button-down collar under a crew neck
  sweater
a street effect: denim jacket and t-shirt
spokespeople: jacket and tie, blouse and jacket
urban or professional: shirt and jacket or turtleneck for men
chic or model: jacket and t-shirt, scoop necks for women
character: loosened tie, or bow tie if you must, but it's grossly overdone

In general, a jacket and shirt opened one button work well for men, a plain
shirt for women. Avoid bold patterns, unironed shirt collars, sweat suits, and
crew neck sweaters with no shirt underneath.

Men with beards, moustaches, five o'clock shadows, or long hair should do
one part of the sitting that way, then shave the beard or trim the hair for the
other part of the session.

Women should color or style their hair within two weeks of the shoot. If
you have several different hairstyles, try changing the style between rolls.

Choosing the Shot

After developing the film from the photo session, the photographer will give you several contact sheets from which you will select the shots
you want enlarged for closer inspection. Included in the package price usually
are two to five enlargements, from which you will make your final choice.
Enlarging more than that will cost approximately $25 extra each. Obviously,
you'll need to use care in choosing these shots. If you are on good terms with
any agents, casting directors, or their assistants, show them your contact sheets
and ask for their input. If you are signed or regularly freelancing with an
agency, your agents will want a say in which ones you choose. If they don't
have the time they will tell you so. Otherwise, they may ask that you drop
them off for a few days. Make sure each sheet is marked with your name and
phone number.

After you've selected your final shot(s), the photographer will retouch it.
Retouching is usually included in the package price. Otherwise, it may cost
$25 per shot or more. Retouching is usually necessary, within limits. Take out
any flaws, but don't erase the character from your face. If you are older, don't
erase all the wrinkles or it will look artificial. Don't retouch anything that

Figure 2.1 RICO ELIAS
Photographer—Nick Granito
Street, ethnic, or just a regular guy

Figure 2.2 NINA TREMBLAY
Photographer—Jinsey Dauk
Engaging, real

Figure 2.3 HANK HADDEN
Photographer—Burt Torchia
Character, blue collar

Figure 2.4 DAVID REIVERS
Photographer—Joe Henson
Young daddy

Figure 2.5 KRISTINA COPELAND
Photographer—Joe Henson
Beauty shot

Figure 2.6 SUZANNE MARSHALL
Photographer—Ron Rinaldi
"Hmmmm...Who is this woman?"

Figure 2.7 MORGAN MARGOLIS
Photographer—Bob Hoeberman
Interesting, young leading man

Figure 2.8 ROBERT STANTON
Photographer—Toni Browning
Young leading man, intense

Figure 2.9 JAMES YOUNG
Photographer—John Hart
Dad, spokesman with an edge, regular
guy

Figure 2.10 RUTH KULERMAN
Photographer—Raffi
Character woman

MARY GORDON MURRAY

Figure 2.11 MARY GORDON MURRAY
Photographer—Lipschis
Interesting, likeable, real

Figure 2.12 GREG CHASE
Spokesman, dad

Figure 2.13 BEN THOMAS
Straight spokesman

Figure 2.14 ANASTASIA TRAINA
Attractive, compelling

Figure 2.15 LAURINE TOWLER
Photographer—Raffi
Straightforward mom shot

Figure 2.16 PETER GUTTMACHER
Photographer—Tom Bloom
"Would you buy a car from this man?"
Character, comedic

Figure 2.17 BEATRICE WELLS
Character, comedic, zany

April Chestner

Figure 2.18 APRIL CHESTNER,
Photographer—Ron Rinaldi
Off-beat, interesting, character

Figure 2.19 PATRICK GARNER, Photographer—Tess Steinkolk
Character, hang-dog quality

Figure 2.20 DAVID McCONEGHEY, Photographer—David M. Brown
Contrasting looks for a postcard

Figure 2.21 APRIL CHESTNER
A non-cliched character shot

APRIL CHESTNER

Figure 2.22 APRIL CHESTNER, Photographer—Marty Wohl
Character composite

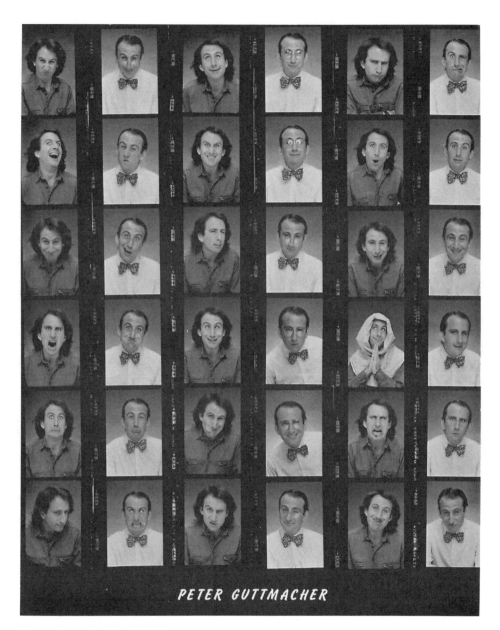

Figure 2.23 PETER GUTTMACHER, Photographer—Tom Bloom
A unique and effective approach

Figure 2.24a LAURA SCRIBNER "Before"

Laura Scríbner

Figure 2.32b LAURA SCRIBNER "After", Photographer—Joe Henson
You can see the difference a capable photographer makes

can't be covered up with make-up or taken off with surgery. You must decide which "imperfections" make you "you" and which are a detraction.

The best headshots give the viewer something in return. Most of the following photographs *work* for their subjects, not by being merely technically proficient, but by revealing something about the person each is portraying. They give an accurate indication of type as well.

Resume

Your resume augments your photograph with enough (but not too much) background material to give the person reading it an idea of who you are and what phase of your career you are in. The idea is to entice without being dishonest. Many actors are tempted to flesh out their experience with euphemisms or invented credits. Don't do this. Firstly, you will be caught in a lie, and when that happens it could do serious damage to your career. Secondly, it simply isn't necessary. Commercially speaking, your face is your ticket, and your experience, training, and skill your method of travel. It is all part of the whole package. Rare is the person who has nothing to offer. If you have no experience but terrific training, then emphasize the training. If you have lots of local theatre experience but no film or television credits, then highlight the theatre work. If you have no training and no experience, then play up your special skills. As mentioned before, there is no such thing as a non-working actor. You should be studying somewhere at all times, and this should always be mentioned on your resume.

Resumes are relatively inexpensive, which is fortunate because they should be continually updated. Paper, so long as it's of relatively sturdy stock, doesn't really matter. Color is up to you as well. Some colored resumes are effective, others obnoxious. One East Coast actress is known for her bright purple resumes. Those who are not fond of them still admit that they are easily located in the file.

Name should be on the middle or side of the top. Do not list your address, but do include both your answering service and home phone number. Many people disagree on this. Some feel you should only include your service number so that unsavory characters will not have access to your home number. While this is indeed a valid point, keep in mind that many times an actor will be contacted at the number appearing on his resume for an interview or audition. Failing to pick up the call from his service in time, he is left to wonder why he wasn't called at his "other" number. One would argue that industry people should have both numbers on file, but this is not always the case. Freelance talent is often called directly from the resume. If you are signed with an agent for both commercial and legit work, you need not include any number other than that of your respective agents. The choice is yours.

STEVEN V. CARR
SAG — AFTRA —AEA

Service: (212) 724-2800   Hair: Light Brown
Home: (212) 384-9870   Eyes: Green
                                  Height: 6'1"

Film

Alphabet City   Policeman
Herbert Hobman, Dir./New Life Cinema

Aliens Among Us   Harry Marcel (Lead)
Ruth Sansonne Dir./TomTom Productions

Television

As The World Turns   Recurring Role   Scott Joseph
All My Children   Day Player
Guiding Light   U/5

Theater

Anything Goes   Sir Evelyn   Entermedia Theater
                                        New York

Kiss Me Kate   Hortincia   Equity Theater
                                        New York

Industrials Have worked for on a regular basis:

Blue Hill Production Co.   Caribeener
Keven Biles Design   Big Apple Production

Commercials Conflicts available upon request

Special Skills

Can juggle 4 balls, English and Irish accents, play tuba, can
catch peanuts in mouth

Figure 2.25a Sample resume

Available for extra work should never be permanently included on a resume. Unless you are making a career out of being an extra, it should be hand written, stamped on, or enclosed in a note.

Commercial resumes should include appearance information such as hair color, eye color, and height. Also provide your social security number. Rather

---

<u>Katherine</u> <u>Adams</u>

Service: (212)840-1600   Hair: Blonde
SS#:   110-36-9847   Eyes: Hazel
                                            Height: 5'8"
AFTRA   Weight: 125 lbs.

<u>Television</u>

   One Life To Live   Becky Sue   Day Player

<u>Commercials</u>

   Available upon request

<u>Training</u>

   Currently studying at HB studios

<u>Skills</u>

   Brown belt in karate, own and operate a motorcycle,
   fluent in sign language, baton twirling, can load
   and shoot a rifle

Figure 2.25b Sample resume

than include an age range, which limits you, leave this for other people to determine.

If you have worked on any commercials, indicate that your *conflicts* are available upon request. These are the commercials you currently appear on. Do not ever list commercials you have done, even if they are off the air, for this will work against you by connecting you to a possible competitor.

Special skills should be included that you do well. If it is a common skill such as baseball, aerobics, or singing, include it only if you are on a professional level, and make note of it. Uncommon skills such as clog dancing, tightrope walking, or parachuting should be mentioned if you are relatively proficient. Don't forget that you may be called upon to display these skills at a moment's notice. If it is a hard-to-find skill, people will be more tolerant. Mention dialects and accents that you speak expertly, identifying which ones you are particularly skilled at. Include musical instruments that you play moderately well, but note if you are on a professional level. Include present and former professions, especially if they can be helpful. Policemen, veterans, and members of the armed forces have been known to book jobs in these capacities because of their true-to-life experiences. Also note if you can drive a stick shift.

The following is an example of a questionnaire that actors may fill out when they sign with an agency. They rate their proficiency at each skill. Referred to by some as a "skills and diseases" list, it is apropos of a business that at times considers age spots and contact lenses an asset. It is included here to give you an idea as to what sort of skills may be mentioned on your resume. Although many of these abilities may never be requested, one just might give you an edge.

Physical Skills/Sports
Acrobatics
Archery
Badminton
Boxing
Fencing
Gymnastics
Jogging
Martial Arts
Mountain Climbing
Running
Stunts
Track and Field
Trampoline
Weight Lifting
Wrestling
Yoga
Baseball
Basketball
Billiards
Bocce
Bowling
Cricket
Croquet
Field Hockey
Football
Golf
Handball
Jai Alai
Lacrosse
Ping Pong
Polo

Racquetball
Soccer
Softball
Squash
Tennis
Volleyball
Bicycling
Drag Racing
Driving
Dune Buggy
Motocross
Motorcycling
Car Racing
Roller Skating
Skate Boarding
Tractor Driving
Unicycling
Boating
Body Surfing
Canoeing
Diving
Kayaking
Rowing
Sailing
Scuba Diving
Skiing
Snorkeling
Surfing
Swimming
Water Polo
Wind Surfing
Figure Skating

Ice Hockey
Ice Skating
Skiing
Snowmobile
Hang Gliding
Parachuting
Pilot
Sky Diving

Horseback
Bareback
English
Jumping
Sidesaddle
Western
Rodeo

Stage/Circus
Aerial
Baton Twirling
Animal Training
Clown
Juggling
Lasso
Magic
Mime
Guns
Knife Throwing
Ventriloquism
Stilts
Elephant Riding

Dance	Woodwinds	Texas
Ballet	Yodel	Scandinavian
Ballroom (specify)	Alto	Continental
Interpretive	Baritone	Yiddish
Jazz	Bass	Irish
Modern	Soprano	Chicago
Square	Tenor	American Indian
Folk (Specify)		East Indian
Tap	*Dialects/Languages*	Scottish
Soft Shoe	Boston	Austrian
Disco	British	Canadian
	European	French Canadian
Music	French	German
Drums	Italian	Middle East
Guitar	New York	Polish
Horn (specify)	Southern	African
Opera	Spanish	Chinese
Piano	New England	Japanese
Violin		

High in demand are expert linguists and dialecticians as well as those who are skilled in whatever particular fad happens to be hot at the moment. In recent years this has included hackey-sacking, bungee jumping, and when *Starlight Express* was on Broadway, roller skating. Once again, only mention skills on your resume that you have mastered or can perform on a professional level, or unusual or trendy skills that you are good at.

Video Reel

Commercial video reels are not necessary for the actor just starting out. They can be invaluable, however, to the professional actor working out of town or tied up on another job, in which case it may win him a job that he is unable to audition or make the callback for. A video reel is handy to show potential commercial agents who are not familiar with your work, but is not mandatory. In fact many of these agents might be just as happy to see your legit reel, should you have one.

An actor should start putting together a reel just as soon as he has professional work to show. For the working actor with no commercials to his credit, quick clips of his film or television work should suffice. Many commercial agents like to have their clients' legit as well as their commercial reels on hand anyway.

Only those actors with professional work on tape should put together a reel. Practice scenes, low-budget showcases, and videos done in class or for the sole purpose of putting together a reel are at best unprofessional, and generally do not show off the performer to an advantage.

Do put together a reel if you have at least one commercial, film, or television appearance to your credit. This does not include extra work. Videos of professional stage presentations or student films should only be included if they are quick, professionally done, show you to an advantage, and are primarily of you. No clip should leave the viewer wondering who the *other* person in the scene was. A hand-held camera version of *A Midsummer Night's Dream* is not particularly advantageous to you as a performer either.

A commercial video reel should be between three and ten minutes in length, and only on the longer end if you have a good many professional commercials to your credit. Versatile actors with a lot of work to show may provide separate tapes, such as one character and one straight reel. Your reel should only contain quality work—no nonunion spots or footage shot by a friend who wants to be a cinematographer some day. Most desirable would be several different commercials and clips of your television and film work showing your many different facets. Do not include old or dated work, as this will only serve to point up how long you've been in the business without becoming a star.

Every time you shoot a commercial, before you leave the set arrange with the producer or production assistant to receive a copy of the finished commercial. Make sure you have the name of the person you spoke to. Be prepared to pay in the neighborhood of $35 for the tape, but occasionally they will get it for you for nothing. Follow up with a phone call one to two weeks later. Do not delegate this to your agent; make it your responsibility to get copies of all your work. If you are not a signed client, it is not a priority for the agent and might not get done. Remember, also, that if you do not arrange to get a copy in advance, you will be hard pressed to obtain one later.

Never under any circumstances give away your master tape. Have copies made immediately—at least one 3/4-inch and one 1/2-inch. Although 3/4 inch is the professional size, many industry people take their tapes home to view and thus prefer 1/2 inch. When your tape is requested, always ask which size is preferred. Put your name and phone number directly on the tape as well as on the box front and spine. This may seem obvious, but any agent can show you her stacks of unclaimed, unlabeled, coverless videos that she is periodically forced to throw away.

Once you have a few commercials under your belt you may feel the urge to have a professional tape made. Although nice to have, it is not necessary. It is an expensive process, often costing several hundred dollars, depending on the amount of editing and effects you want to include. If you are at the stage of your career where you feel it would be advantageous to have such a tape, you can keep the cost down by doing a lot of the preparation work yourself. Pinpoint the order and where the scenes should be cut beforehand. Like your resume, your reel should be continually updated. Getting rid of the old and putting on the new can also be costly. Try to work out a price in advance for re-editing later on.

If you opt for a professional reel, the opening shot should be of your name, phone number, agency and phone number if applicable, followed by a still of your headshot. Slow fade-ins, fade-outs and other special effects are distracting, time consuming, and, as mentioned before, costly. Keep it simple, quick, and clean. You may end the reel with another shot of your name and number, but inserting a final look at your headshot is unnecessary.

You and your agent should each have at least one reel of your work in each size. The master should be kept in a safe and readily accessible place. Some actors have been known to keep their master tapes back in their hometown with Mom and Dad, which doesn't help when a copy needs to be made in a hurry. Videos are often requested at the last minute and must be delivered to the ad agency or expressed to a client with little advance notice. There is rarely time for eleventh hour editing or last minute scrambling for copies. Keep a log of the location of your reels, and check up on them when they are not returned promptly.

In short, if you have professional, quality work to your credit put together a reel. No reel is better than a bad reel. If you can afford it or your position within the industry warrants it, have one professionally made. Always keep your video up-to-date, with plenty of copies on hand.

The Inner You

It is up to the actor to create and radiate his own persona. It is entirely up to him as to whether he will be perceived as a winner or a loser. Hopefully this chapter has given you the basics from which you will perfect most aspects of your product:

attitude
technique
appearance
selling tools—pictures, resumes and video
knowledge of type
knowledge of the industry, which will be gained from further reading of this
    book, networking, and and just plain experiencing

The one thing that cannot be taught is *self-knowledge*. One cannot find a job, much less carve out a career in this business, without a clear understanding of who he is, both inside and out. Sometimes this happens at a very tender age, sometimes later in life, and sometimes never. When self-knowledge happens, something "clicks" and all the elements seem to fall into place. It is at this point that an actor may really start to work. Until that happens all he can do is study, make contacts, perfect his product, and persevere. Most of all it is important to keep active and constantly working, in whatever capacity.

In addition, one of the smartest things an actor can do is to develop outside interests. Find a job or side career that keeps you interested and stimulated. Being with people other than show business types can keep your perspective fresh and your spirits up. A fulfilling job not only makes for the well rounded individual, but having a fall-back position will help take the pressure off when you audition. Once you realize that you don't *have* to have a particular acting job, something may break for you. A side career can make the difference between being desperate and unhirable, and being mentally healthy and working in commercials.

Interview with Joan See
Instructor/Director
Actors in Advertising
AIA/Three Of Us
39 W. 19 St.
New York, NY 10036

BEARDSLEY: What kind of classes do you give at your school?

SEE: Our classes range from a basic commercial class for someone who has no training to a two-year acting program for film and television based in Meissner acting training.

BEARDSLEY: How did you get into this field?

SEE: I have a degree in English Speaking Theatre and have been an actress for my entire life. I was trained by Sandy Meisner and Wynn Handman, and have acted in countless commercials. I have been an actress for more than 30 years.

BEARDSLEY: Talking specifically about the commercial division, what makes your classes different?

SEE: They are based on the premise that there is no such thing as commercial acting. There *is* such a thing as acting in commercials, as there is acting in soap operas and acting in feature films. The first thing that makes commercials different for any person who is familiar with the stage is that a camera is present. We are acting training based, technique based, skill based. All of the skill and technique classes are taught by actors who have been successfully doing what they teach, and also happen to be excellent teachers with excellent backgrounds in teaching. We don't bring casting directors in as teachers until the talent is prepared to show off well. We don't have casting directors teaching basic skills technique classes. Everyone moves through the program at their own speed, depending upon their skill level and upon what problems they have with their basic acting technique. If somebody has a vocal problem or a very serious articulation problem, they are going to require more than a class in acting in commercials. We do a great deal of evaluation and counseling. We would certainly counsel those students, if they wish to be marketable and to get a positive response from agents and casting directors, on what they need to work on. We would encourage them to do it with our faculty or we would, in some cases, recommend them to other people who are more appropriate, depending upon the problem.

BEARDSLEY: How do you screen for enrollment?

SEE: We interview everyone who wants to work with us. We bring them in, talk to them, and in some cases have them read copy. We look at resumes, we talk about their background. We then insist that they audit what we think is an appropriate class and have them evaluate if it is appropriate for them. Placement of the student is based upon their skill level and experience. They do not move through a program by rote.

BEARDSLEY: Do you turn down people who are just not skilled enough or together enough to make it commercially? Can you even make that kind of judgment?

SEE: Anybody can be taught the skills. Whether they have the imagination and the courage to take the skills to the level that we would hope is up to them. Imagination is something you can't give to someone. That's what I call talent. Someone with very good skills training, moderate talent, and good marketing ability can do just as well as someone who has a lot of talent, no technique, no marketing ability, and no idea what they are doing in the business part of the business. The mark of success, I think, in on-camera commercials specifically, is not necessarily relative to a person's degree of skill mastery. When you are dealing with on-camera, you are dealing with fashion in terms of who is castable now. People who are castable now would never have been castable 15 years ago. One of my most successful students can't book fast enough. Fifteen years ago a commercial agent would have looked at her and said, "You have got to be kidding." She is long, lanky, and quirky, but right now she is what everybody is looking for. There are many subtleties and complexities in answering that question. To be successful, actors must get a lot of things right—they must have the skills, the talent, be the right type for this particular period of time, be well organized, and have good, outgoing, businesslike personalities. Then they must get a break. Any one of those things being out of sync could blow the whole thing out of the water.

BEARDSLEY: How do you define acting in commercials and what are the common misconceptions?

SEE: It is communication of where you are that does not necessarily equate to smaller. People who are being taught that smaller is what film acting is about wind up being stiff and without energy. They end up being talking heads because smaller, smaller, smaller, means don't move, don't move, don't move, don't move, don't move. Anybody who wants to talk about film acting being smaller should go to a Jack Nicholson film festival. "Smaller" is perhaps the most toxic idea about acting for the camera that is currently bantered around in teaching circles. Commercials in the 1990s are mini-movies. They demand that there be an understanding of film acting. The camera is the first thing that has to be mastered. Any actor from stage, or who has dealt only in scene classes in New York, comes from a whole different mindset and group of skills. The camera doesn't change what acting is about. It changes what you think about acting. It is different, but it is not smaller.

BEARDSLEY: Then the most frequent problem you come across is that the actors are trying to tone it down for the camera?

SEE: The thing that I come across most often is basic lack of ability to concentrate, followed by the erroneous perception that acting is *feeling*. Acting is *behaving*. In a lot of their training, actors focus on feelings, feelings, feelings and then don't appropriately go beyond that. I find a lack of a real understanding about d*oing*, about *simplicity*, and about *concentration*. Most actors that I work with just don't know how to *be* and think in front of people. Until you know how to be and think in front of people instead of do and emote, you cannot do film acting of any kind.

BEARDSLEY: Who is easier to work with, trained actors or someone new off the street who has an instinct?

SEE: That is a very individual call. Everybody is very different. Talented, motivated beginners can be a joy because they have no preconceived notions and they're like writing on a clean piece of paper. As long as you keep faith with them they will flourish and do very well. Then again, very talented actors who have solid technique and understand that technique, and what they are doing, can be just as much of a joy. They go back to their technique, and it all makes sense to them. But the basic requirement is having a fairly unflawed technical instrument—with little or no blockage either physically or vocally.

BEARDSLEY: And how do you define that blockage?

SEE: They are not physically tight, they know how to breathe, their voice is centered, they have control of their body, they have control of their voice. They know how to control their anxiety through breathing and concentration. They know how to handle simple things such as preparation, actions, and objectives.

BEARDSLEY: Can acting in commercials assist the actor with his technique?

SEE: Absolutely. A compressed text, which is what I call a commercial, gives a character 28 seconds, whereas a character in a soap opera can live for 28 years. What's needed in that 28-second story is speed, clarity, and believability. So one of the things about working with good actors in this medium is that you can read through technique. A lot of junk can be cleared out of the way. They can walk out of a class a better actor in all respects. After I had been working primarily in commercials for a while, then went back and acted in front of an audience in theatre, I discovered that it was easier. Somehow it was simplified. I began to realize that the demands that were put on me in 28 seconds were forcing me not to be self indulgent. To be very clear, to use my technique in a way that was going to produce a very specific result, not a general result. There is one bad habit that commercials will absolutely break in a minute.

BEARDSLEY: Which is?

SEE: Thinking that generalities are safe. Commercials are about specificity and details, and the more specific you are and the more detailed you are, the more you think about all that, and the more you deal with that before you go into the room, the more likely you are to get a callback. After that, all bets are off— as you well know.

BEARDSLEY: Are there any skill areas that aspiring commercial actors can improve on their own?

SEE: One of the most pervasive problems among all actors is the depth of their vocal problems. Vocal problems are not addressed by most schools because they are not sexy and they are hard to sell. It is hard to get people into voice production classes. Take it from an administrative staff sitting here next to me shaking their heads, "Yes, yes, yes." Voices are not being trained correctly in America.

BEARDSLEY: Are you talking in terms of tonal quality?

SEE: Speaking voices, tonal quality, breathing, and tone. A voice that is supported. A voice that is not a little girl's voice way up in her head. Not breathy, the voice of a grown-up. A voice that is controlled, not grating, a voice that is in the mid-range and orally resonated, is what I wanted in commercials. The same thing is necessary for the theatre and for soap operas. The depth of vocal problems, even in working actors today in New York City, is monumental. Most teachers do not want to bring it up. Actors all say, "I know, I have been told that forever." We respond by saying, "Then I am not putting you in front of agents and casting directors until you fix it." Because it is not appropriate. Most actors have to stop looking for quick fixes, take care of their pianos— which are their bodies and their voices—and stop looking for people to make magic. This is a serious business. Commercials are serious business. This is not something you do in your spare time and knock off because somebody told you you were cute and you ought to be doing it. It is a multimillion-dollar industry. More actors make money from commercials than any other venue, and if they want their share of the pie, they better start paying attention. Serious attention. They will go and spend mega, mega bucks for a third rate acting teacher and they will not think twice about spending no dollars in a school like this in order to get their commercial technique together. Yet they want the residuals. You can't have it both ways.

3

□ □ □
□ □ □
□ □ □

Marketing Yourself

Once you have packaged the product, you must learn how to market it. Success is determined not by merely having something to offer, but by knowing what to do with it once you have it. Whether you are on the East or West Coast, finding an agent should be your first order of business. In New York you may seek work on your own behalf or work with one or several agents. In Los Angeles you must be sent out exclusively by one agent or agency. Whether you are looking to sign or freelance, working with and through an agent will open doors for you, give you more stature, and hopefully make your life a little easier. This will be explored more fully in Chapter 4.

Making the Rounds Count

There are many ways of meeting an agent, some better than others. If you are sent to an agent by someone she respects you have a much better chance of getting an interview than if you corner her in an elevator and force her to listen to a rendition of all your favorite character voices. While if you are very, very good this might work, in all probability it will simply cause her to use the stairs in the future. Acceptable ways of meeting an agent, placed loosely in order of desirability are as follows:

1. Through the recommendation of a casting director, director, or some other person of stature within the industry who knows and respects your work
2. Being approached personally by the agent, as the result of her seeing your work
3. Being called in because you booked a job over one of her other clients
4. Being brought in as a result of the agent's assistant seeing your work
5. As a favor to a client or someone else the agent wants to please
6. Through a picture and resume
7. By approaching an agent on the street, elevator or anyplace outside her office

Keep in mind the psychology of being the pursued verses the pursuer: The person doing the pursuing sees the object of his aspiration as being more desirable than if he were being pursued by it. Getting an agent to want you is half the battle. In the first five examples, the agent is doing some degree of pursuing. In the last two you are the pursuer. Obtaining an agent is a marriage dance, and you are now trying to get someone to date you.

Mailings

While pictures and resumes may not always directly work to get you an appointment, sending them out is still a vital part of the process. If you are in New York, get yourself a *Ross Reports*, which can be obtained through drama or campus bookstores, your local AFTRA office, or directly through Television Index, Inc. at 40-29 27th St., Long Island City, NY 11101. It is very inexpensive and is issued monthly. It contains updated lists of casting directors, agents, and advertising agencies, designates which cast commercials, who accepts visits or phone calls, and so on. It is indispensable to the beginning East Coast actor and should be bought often to keep the reader abreast of industry personnel changes. Los Angeles actors do not have an equivalent publication, but lists of agents and casting directors can be obtained in drama bookstores and union offices. Take care in checking spellings and employee changes before addressing your mailing labels.

Recognition Factor

Some people stand a good chance of being called in from their picture and resume, particularly:

- good looking men age 30 and above
- average to good looking men who can act
- professional models, male or female, of any age
- actors with good acting credits, or who have done several commercials
- mid-American ethnic types, particularly in the older age ranges
- those possessing a skill that is currently "hot"

Or, you may hit an agent on a good day, you may have gone to the same high school as the agent's assistant's brother's girlfriend, she may have just taken a breakdown for your exact type, or she may just like your face. Regardless of whether any of these factors can be attributed to you, getting your picture in front of an agent as frequently as possible helps heighten your *recognition factor*. An agent or casting director receives as many as 50 photos a day, interviews two to ten actors a day or auditions 100, may talk to 30 people on the phone, and then goes to see a show with a cast of 25. Getting your face out there as much as possible, for as long as possible, is vital. The more your face is seen, the higher your recognition factor. Sending out pictures and resumes in and of themselves may not get you an interview, but they do contribute to this end result.

The Cover Letter

Send your picture and resume to as many appropriate people as possible. Your initial mailing should be to all agents and casting directors who handle commercials. Make sure to get correct spellings of names. Always personalize each cover letter; never use general salutations such as "Dear Sir or Madam" or "To Whom it May Concern." Know whom you are addressing; letters sent to the improper party may end up in the dead file, more appropriately called the garbage. If you are in doubt, call the receptionist and get the correct information. Women should be addressed as "Ms.", men as "Mr." Do not address anyone by first name unless you are already on a first-name basis.

The letter should be neatly typed, with typos whitened out, not crossed out. It is better not to hand write unless your writing is *extremely* legible. You can forego expensive or fancy stationery for simpler stock, as long as it is not looseleaf paper or a page torn from a notebook. Good quality typewriter bond is inexpensive, readily available, and acceptable. Your letters should be neat, clean, brief, and to the point.

Do your homework. Through researching and networking you should know a little about each agency and casting office. Some agencies are known for their legit actors. Others primarily use models. Still others may be known for having actors with hard-to-find skills. Some casting directors cast only for extras or do primarily minority casting. By knowing which buttons to push, you can develop an edge over your competition.

Key phrases that get an agent or casting director's attention are "I booked such and such job," "got called back on...," "got a first refusal on...," "so-and-so (respected industry person) knows/likes my work," "I just finished doing work on...," "I'm being sent out by...," and so forth. In fact, nothing perks an agent's ears up more than hearing that you are being sent out by *her* competition. They are only human, after all.

Let the agent or casting director know if you are new in town, what work you currently may be doing, and which commercials you may have booked or have gotten callbacks on. Let agents know which casting directors are familiar with your work, let casting directors know which agents are sending you out. If you are contacting them at the recommendation of someone else, tell them so. Generally speaking, give enough information to be enticing without giving your life story. Screen Actors Guild members: Note in the letter if you are available for extra work. If you are promoting a special skill, for instance if you are a member of the "Ice Capades" and would like to audition while you are working in town, note that fact in your cover letter as well. In either of these cases, enclose an extra picture and resume, as agents and casting directors often have files for extras and special skills. Indicate in the letter if you have a videotape, but don't send one unsolicited.

Sample Cover Letters

Dear Ms. Anderson:

Although I have been successfully freelancing for the past year with the Milrose Agency and Harriet Carter, I feel it's time to sign exclusively with one agency.

Currently, I have two spots on the air, a local Tyson's Chicken and a national Slim-Fast.

I would be happy to forward my commercial reel for your review, or come in and talk with you in person.

Sincerely,

Katherine Adams

Katherine Adams

Dear Ms. Anderson:

Your client, Patrick Goodman, suggested I contact you. We worked together on *The Cherry Orchard* at Seattle Rep. Enclosed is a review of my work from that production.

Although I have not done commercials I have been told by casting directors that I would do well in this area.

I will be back in town on the 22nd of this month, and look forward to meeting with you after then.

Sincerely,

Steven Carr

Steven Carr

Some commercial instructors tell their students to put in the body of the letter that they'll be calling at a certain time to set up a meeting. This is presumptuous on the part of the actor, and, given the volume of mail received by an agent, most definitely will be forgotten.

Going Out in Person

In addition to regular mailings you should be making the rounds in person. Although this is discouraged by many offices, you will learn quickly which those are. Make it a point to visit as many offices as possible on days you are not working. Most likely you will not be seen by anyone other than the receptionist, but you would be surprised to find out how many times this person is actually the casting director or agent. In small offices, especially, one person wears many hats. Although they may not own up to it, the person you are handing your photo to may have more clout than you realize. On the other hand, don't overlook the importance of office assistants. Often the agent or casting director will see people at their recommendation. If you are a type they may be in need of that day, or if they like your smile or look, you could be in luck. Some places have glass partitions or open office areas which enable anyone walking in to be seen by the whole office. In my office, which had just such a partition, agents and assistants frequently spoke to walk-in actors who caught their eye. However the office may be set up, politely ask the front desk person if you could be seen or read copy for anyone there. Even if you are turned down you will have gotten the opportunity to have a friendly word with at least one person.

Whether speaking to people by phone or in person, always remember to jot down pertinent information. Besides recording general information, note snippets of conversation that can be brought up in future interchanges.

Follow-up

Do not follow up your initial 8x10 mailing with another one right away. Instead, use post cards or show flyers to generate a response. Three to six months later you may try another mailing with a different photograph.

Postcards

Follow up your first mailing (of photo and resume) with a postcard a month later. Postcards should be brief and positive. By letting the reader know what you have been doing since you were last in contact, they serve as a subtle reminder that you are still available.

For example:

Dear Ms. Anderson:

   Since we last spoke I have received first refusals for a McDonalds spot and a Reeses Pieces. I have
an interview to meet with The ABC Agency next week.

                              —Katherine Adams

Not:

"Why haven't you called?," or "Hi! Remember me???"

Phone Calls

   Some industry people recommend following up your mailings with a phone call, others vehemently oppose this practice. There is no clear answer. Most offices are too busy to take unsolicited phone calls from actors and will tell you so. However, you may be lucky and get an agent or assistant on the phone who is in a beneficent mood with a moment to talk. If this should occur, use the opportunity to your advantage. It does not happen often and should not be wasted with indecision. Ask that person what it takes to be seen by an agent in their office, what types they may be in need of, or invite them to your show. Invite that assistant out for lunch or a cup of coffee. Overworked and underpaid assistants may be happy to accept offers of free food. Do not try to engage them in idle chit-chat or badger them with stupid questions. Thank them for their time. Also, keep in mind that many times agents or casting directors answer their assistant's phone and do not let on who they really are. The following exchange happened one day in our office:

Agent: "Hello, Genie's line."
Actor: "Can I speak to (agent who he is actually speaking to)...?
Agent: "Why?"
Actor: "I wanted to see if she received my picture."
Agent: "Do you have any idea how many pictures an agent in this office receives in a single day?"
Actor: "Uh..."
Agent: "What makes you think she would remember yours?"
Actor: "Uh..."

You can see what you may be in for. Be ready with intelligent questions—and answers.

Keeping Records

Keep a log of the people you've contacted. If you have been sending them material for several months with no response, you may be addressing it to the wrong party, or it's time to back off and try another tactic.

Being Seen in Your Showcase

The best and most lasting impressions are made when an actor's work is seen. While film, television, and professional theatre are the best ways to get seen, showcases are also an acceptable and accessible method of getting work observed. These are usually low-budget productions, put on with minimal sets and scenery in less than sumptuous surroundings. The talent and production staff usually participate for the sole purpose of gaining experience and exposure. Some showcases are extremely well done and are viable avenues for those seeking to display or find new talent, while others are embarrassing fiascos with too little attention paid to rehearsal, choice of quality actors, directors, written material, and surroundings. The space need not be luxurious, but neither should the toilet be in the audience area separated only by a thin partition. In order to insure that his time is not wasted, the actor wishing to be seen in the showcase must make sure that the show is of relatively high quality and will be attended by industry professionals. Also, he must be sure that his own work is ready to be seen, or it will work against him.

The following checklist will help make showcases attractive to industry people. While there are no guarantees, using some or all of these suggestions will help increase your odds of getting a positive return for your efforts.

1. Find out which shows are being run by respected groups and which ones have a reputation for using bad talent. If it is a start-up production that you are putting together with friends, make sure the director is accomplished.
2. The level of talent in the production should be equal to yours or better. You want to be involved with a group of people who reflect well on you and who attract an audience that can help you. Actors who are signed to an agency are likely to bring in their agents to view their work in the show. In fact, entire agencies have been known to come support their client in an endeavor.
3. The work being done should be tried and true but not overdone. If it is a new play, read the script with discernment. It's awfully hard to look good in a bad play.
4. The location should be in a decent neighborhood not too far off the beaten path. If your play happens to be less than conveniently located, you might want to get together with the other actors and hire a car service to ferry agents to and from the show. This may sound drastic,

but it may be the only way to insure that the proper people will be in attendance.

5. Take agents' and casting directors' schedules into account when planning the time and length of the production. If it is a high-budgeted production, the agents may bring a casting director as a guest, and will want to have dinner first. Top-notch productions may start at 8:00 P.M. On the West Coast some agents cover shows at lunchtime. Low-budget and untried arenas, and as most showcase productions are, will generally attract agents and assistants coming directly from work, and would do best by starting directly after office hours, at about 6:30 or 7:00 P.M. Running length for these shows should be from 60 to 90 minutes, with no intermission. If the people you are trying to attract have to wait around till 8:00 P.M. for your show to start, they may very well go home instead. If they know they can cover a show and still be home at a reasonable hour, they just may choose to attend your showcase over that of someone else. Schedule performances Mondays to Thursdays, as many people reserve the weekends for major theatre or personal pursuits.

6. The larger the cast, the more attractive it is to agents. Remember, they are covering shows to find new talent. Don't go overboard, but if an agent has a choice between covering two people or eight people in a single evening, she is likely to go for the larger of the two groups.

7. If it is summertime, make sure the performance space has air conditioning, or you will find yourself playing to a room full of empty seats.

8. Contrary to popular belief, wine and cheese, while a nice touch, does not serve to entice anyone other than starving actors to your show.

9. Packets containing photos and resumes of the cast should be available at the door before the performance.

10. Each actor should take it upon himself to personally invite agents and casting directors to the show. *Do not* leave it up to the production staff. This is a good excuse to hand deliver flyers to offices. Do not overlook assistants. They are often the ones covering showcases. Take the time and care to determine the names and correct spellings of the assistants in each office. As they do not always receive personal invitations, this may very well attract them to your show.

Talent Directories

You may be confronted with a myriad of different organizations trying to sell you their particular method of getting your face in front of industry people. For a fee, they will include your picture in mailings, video reels, magazines, newsletters, cable shows, computer transmissions, and other new and as-yet-unheard-of ways. Some of these services are blatant excuses to take your money. (See the section on money matters in Chapter 7.)

There are only a handful of tried-and-true talent directories. Two industry standards are:

The Player's Guide
1500 Broadway
New York, NY 10036

Academy Players Directory
Academy of Motion Picture Arts and Sciences
8949 Wilshire Blvd.
Beverly Hills, CA 90211

Listing yourself in these guides is well worth the cost. They are routinely used by casting directors, agents, producers, and directors for reference and new ideas. They are updated and reissued annually. Although you may never know it, being in the appropriate guide may be directly or indirectly responsible for your getting a particular audition or job.

Be bold and innovative when trying to get in to see an agent or showcasing your talent. Do not be pushy or obnoxious. If one method doesn't work, try another. One group of singer/actors puts on a marvelous *a capella* medley of Christmas carols, taking it around to offices at holiday time. Not only does it showcase their talent, but it proves a delightful respite for industry people in need of a break. Because it is original, bold without being overbearing, and well done, it leaves a lasting and positive impression.

Take the needs of the people you are trying to reach into consideration when planning your moves. If it can serve both your purpose and theirs, you may very well steer some good fortune your way.

Interview with Don Snell
Marketing Director/Actor
Corporate Productions, Inc.
4516 Mariota Ave.
Taluca Lake, CA 91602

BEARDSLEY: What do you do as marketing director for Corporate Productions?
SNELL: I find clients who wish to produce industrial films. I devise the marketing strategy and develop marketing tools such as capability brochures and demo reels. I am also an actor.

BEARDSLEY: Do you have any marketing information for actors?

SNELL: Yes. Be able to say your name out loud. It is so simple. This is a telemarketing business supported by direct mail.

When a casting director needs an actor, he likely will call five agents and ask for five people from each. When asked for five actors an agent will recommend ten. The casting director will jot down those names or receive them via fax. Name recognition gives any actor an edge. Even if your name is Sigourney Weaver or Dabney Coleman, the casting director must recognize and be able to pronounce it or else he can't say, "I would like to have Sigourney Weaver come in."

BEARDSLEY: And they learn to say it out loud by seeing you on the premises?

SNELL: Correct. When I first came to New York as an actor, I sent everyone *one-sheets*. These were single sheets of paper containing my name, photograph, and a brief statement about me, my type, my work activities, and a phone number. With those four elements: who I was, what I was, what I was doing, and how to reach me, I was able to imprint the identity of Don Snell. In a very short time casting directors knew who I was, and it was very easy for them to say "yes" or "no." I did not waste their time and they did not waste mine.

Another thing to remember is that you never know who is talking about you out there. Here is a case in point: You check your service on Monday and there is no message. You go through the whole week with no message. On Friday there is a message to call your agent. Now that means that somebody was talking about you probably on Thursday and maybe as far back as Wednesday; and it could have been that on Monday when you were so depressed because there was no message, your agent was putting your 8" x 10" photo in an envelope and mailing it. When I got into this field I made the commitment to do *something every day* towards it. May I give you my little mathematics now?

BEARDSLEY: Yes, please.

SNELL: Write down the number 365. Underneath that write 240. Now subtract that and you have 125. Most people work 240 days a year. They work five days a week and they take holidays and a vacation off. Still, there are 365 days available to us. Those 125 days can make your career. I worked on *Places In the Heart* with Sally Field. She made that movie in 65 days. Jennifer Jones won an Academy Award for one 5-minute scene. She probably shot it in two days. So it does not matter in your career how many days. You can work two and win an Academy Award. You can work 65 and win an Academy Award.

Now go back to the 365 and add two zeros to it. If you make $100 a day, you are making a living; you can pay your rent, buy food, and have some clothes to wear to your auditions. If you set that manageable goal of $100 a day, you have $36,500.

Now add another zero to that number, and you have your first year as a contract player on "All My Children." Add another zero, and it's your first year as a Tim Allen on "Home Improvement." Add another zero and that is your tenth year as Oprah Winfrey.

These people are doing what they love to do and are very well compensated for it. Oprah Winfrey has been nominated for an Oscar and has a talk show and her own production company. There are no role models anymore. We create our own. I work as a marketing director for a production company, but I will still take time to audition for a casting director. If I get the job, I fly off and do the job.

I just returned from shooting a commercial in Denver. As soon as I got my production dates, I started calling on clients in Denver and letting them know that I was going to be in the area and would like to visit with them. I shot the commercial and stayed two more days to visit clients.

BEARDSLEY: Not a second wasted nor an opportunity lost.

SNELL: It's all part of being in business. As the song says, "There's no *business*, like show *business*, like no *business* I know." Another marketing technique I recommend is to have a list of sources of income. I have prioritized my list not by the amount of money that can be generated but by artistic or career goals. We would all like to be like Sally Field and go from being Gidget to winning two Oscars, which puts her in a league with Vivian Leigh, but few actors have done that. So at the top of my list is major motion pictures, followed by a movie of the week, which is shot in the same way as a film. Under that is prime time television, then daytime television; and number five for me is theatre. For some people, theatre tops the list. Then I have commercial television, commercial radio, industrials, voiceover work, and commercial print.

I have made anywhere from $500 to $5,000 a year on commercial print. I have one friend whose photo is on a product box, and he earns $10,000 a year just from that. I did a commercial for Nyquil™, and out of nowhere they asked what my fee would be for the box. Nick Nolte used to be on the box for Summer Blonde™. Corbin Bernsen of "L.A. Law" is on a point-of-purchase display for pocket combs that looks like it was shot in 1973.

By creating that list, I remind myself where I want to focus my efforts. You can make a whole lot more money from one day on a commercial than you can in 63 days on a film sometimes. First decide what you want, what's more important to you.

BEARDSLEY: What is your philosophy on this business?

SNELL: When God closes the door, you find out the windows are painted shut. So pick up a chair and throw it through the window. I'm one who is called back often and put on hold even more, which can be exhilarating but also depressing when they call and say, "Oh, we have to release you." So when I am put on hold, I go to that date in the appointment book and write, "Hold" with (the client's name). When they call and release me, I go back and make a simple X through the entry, making sure it is still legible. That way, when that date rolls around and I am not on the set shooting the commercial, I see what happens to me that day that couldn't have happened if I were doing the commercial. I only use a pencil in my appointment book so that my schedule is adjustable.

BEARDSLEY: How did you make the move from the East to West Coast?

SNELL: I did a student film at NYU that won an award out here and was shown at the Director's Guild. It aired on PBS in January 1992. By doing that film I not only got my job at the production company but I got an agent here as well, moved, and the rest, as they say, is history.

BEARDSLEY: Did you have trouble finding an agent?

SNELL: Beverly Hecht became my television/film agent after I mailed 180 birthday announcements. She called and you could hear her laughing. The announcement said, "Date of Birth: May 30, 1949, New Category: 40–45, No Gifts Please." All I needed was one person to call, and she did. When I made the decision to move here in January 1991, I sent an annual report to 180 agents listing everyone I worked for in 1990, a phone number, and a picture. I received a response from about eight agents.

BEARDSLEY: What do you think is a good response percentage?

SNELL: Catalog houses consider 3% very good. Most direct mail people are very happy with 1%.

BEARDSLEY: What do *you* think is good?

SNELL: The direct response is usually measurable because somebody picks up the phone and calls you. But the indirect response is what I basically work for. Over the course of the year everyone on my mailing list sees my picture and name maybe eight or nine times so that when my agent submits me for another project, they look at my name and know it, but they don't know why. They do, however, know how to say Don Snell.

BEARDSLEY: Do you feel there is any difference between New York and Los Angeles in the marketing area? Do you use different techniques to generate a response?

SNELL: Something about New York is so much easier. I think it is because everyone is captive there on that little island. We're all real tight, and we are all up against the same elements for some reason. I don't mean to disparage Los Angeles, because I live here. The difference out here is that it is slower. It doesn't surprise me when someone receives my mailing and wants to respond but doesn't contact me until two weeks later. There is such an immediacy about New York. It has to happen *today*.

I plan to be here for a long time. I moved to New York to move out here. I know that is a little roundabout, but I had sold a business and did not want to move here as an ex-manufacturer, but as an actor. I was able to do that by going to New York and studying and working in front of the camera. That way I was able to arrive here with video footage.

BEARDSLEY: Is there anything a performer can do to be more salable in the industrial marketplace?

SNELL: Learn how to use a tape recorder. This is a plus, especially for on-camera spokespeople. I am referring to an instrument called an *ear wig*, or ear device, which is a tiny tape recorder you use to record your copy, then play back in your ear. The small clear plastic ones are just like the ones that television journalists use. You can't see them on camera. They cost about $200. It is a thing you need to practice with so that you become proficient at it, like any other learned skill.

BEARDSLEY: Do the industrial companies prefer an ear device or teleprompters?

SNELL: Knowing how to use an ear device gives a company one more reason to use you. If you are going after spokesperson work, it is an excellent skill to have and should be on your industrial resume, nicely blocked off so that it gets the reader's attention.

BEARDSLEY: Are there any other eye-catchers on an industrial resume?

SNELL: Just remember that it is acting work in a business context. If you have a business background at all, which many actors do, include it there. With computers, resumes are so easy to adapt for the application. I keep five or six different versions of mine on a disk.

There is one more thing I wanted to mention about industrials. Price Waterhouse has an in-house facility, and when you work there they take your 8" x 10" and reduce it down to 4" x 5" and put it up on the wall. They put everybody up there . . . Ted Danson is up there. When I worked there I saw a number of faces I recognized, such as Bruce Jenner, and Michele Nichols from *Star Trek*. We just shot an industrial with Walter Cronkite. John Cleese is one of the biggest in industrials. We worked a lot out here with Richard Anderson who played in "The Six Million Dollar Man." I was at one of the advertising agencies and saw a piece of paper that looked like a call sheet with a list of actors names. I realized it continued voiceover possibilities. The names on it included William Shatner—the big boys. They get really nice sizeable fees for these kinds of jobs. Michael Douglas and Donald Sutherland are doing voiceover work right now for a series of commercials. When I saw this list, I realized, you are still auditioning, even though you're William Shatner.

4

□ □ □
□ □ □
□ □ □

Finding, Getting, and Keeping an Agent

Most actors are surprised to learn that agents are usually very ordinary people with their own needs and wants, shortcomings and strengths, and their own individual ways of dealing within the industry. Be they admired, feared, disdained, or loved, it is generally agreed that they are quite necessary. There is no black-and-white resolution to any issue, and nowhere is that more apparent than in the pursuit and selection of an agent.

What an Agent Does

The Screen Actors Guild (SAG) defines a *talent agent* as:

A person, copartnership, association, firm or corporation who or which offers to or does represent, act as the representative of, negotiate for, procure employment for, counsel or advise any member of the SAG in and about and in connection with or relating to his employment or professional career.

A *sub-agent* is a person who is employed by a franchised agent to function in the same capacity as noted above.

A *commercial agent's* duties and services are delineated as follows:

1) *Seek and arrange interviews, negotiate terms and conditions of employment, and examine proposed employment contracts to check conformity with deal negotiated;*
2) *Advise actor concerning any provisions of the employment contract pertaining to exclusivity, releases, warranties or other special clauses;*
3) *Maintain records and keep actor advised of any exclusivity commitments, use best efforts to clear conflicting exclusivity commitments and engagements and obtain releases for actor where necessary, negotiate for releases of exclusivity commitments and other restrictions where commercials have been withdrawn from use;*

4) *Maintain adequate records showing dates of employment, dates of first usage, class of usage, cycles of usage, and payments made for employment and usage;*

5) *Where necessary, send reminder to employer of payments due for employment and usage and promptly report to SAG any cases of repeated late payments and other violations;*

6) *Where employer seeks to acquire other rights or services in addition to the performance of the actor in a commercial, agent shall bargain separately for such rights and services;*

7) *Maintain records regarding maximum periods of use and reuse, advise actor of expiration dates of periods of use, give written notices to advertising agencies of actor's election not to grant right of renewed use;*

8) *Make periodic inquiries to determine if commercials have been withdrawn from use;*

9) *With respect to compensation for television commercials collected by the agent and paid over to the actor: the agent shall accompany each such check with a voucher which shall contain the name of the employer or advertising agency, name of product, nature of payment, cycle dates and date of payment.*

10) *Notify the actor and SAG whenever a late payment is due an actor.*

11) *The agent shall notify the actor not less than 120 days prior to the expiration of the maximum period of use of the forthcoming expiration of said period.*

But what does an agent *really* do?
An agent will

send you out on auditions
negotiate your contracts
help you choose a photographer and photographs
recommend a number of appropriate acting coaches or commercial classes
come see you in your shows
take your phone calls
reschedule audition times for you and make sure you get in, within reason
make sure all the casting directors know you
help you find out why you are not booking
reassure you when you are not booking
help you sift and select jobs when the time is right
keep your contracts, conflicts, and first refusals in order

A *good* agent will

be one of the first to be called on every job
have the respect of casting directors

offer you new insights into yourself and your place in the business

be pleasant and informative when you call or come by

bring industry people who don't know your work to your shows

call you regularly to keep in touch, particularly when business has been slow

ask for and sometimes get feedback for you

help refine and direct your image and talents

recognize the importance of your legit work and understand when it has to take precedence

not release you when you are not booking

smooth over differences between you and casting directors

try to insure "favored nation" status (that no one makes more money than you) whenever possible

show imagination and creativity when submitting you for jobs

be thorough and knowledgeable when giving out audition and booking information

make sure you are well compensated for extras such as print, cable, or foreign usage

not be bullied by casting directors, but get you what is rightfully yours

A *fantastic* agent will

get all the calls and get them first

always make sure you are the highest paid actor on the set

move Heaven and Earth to enable you to do a job you thought you had to pass on

be farsighted and keep your entire career in mind, not just the foreseeable future

believe in you no matter what

An agent will *not*

get you the job

continue working for you if you repeatedly do not show up, show up late, turn down auditions for no good reason, or don't give your all

A *bad* agent

is unimaginative, unable to grow when you are

gives out erroneous or incomplete booking or audition information

sends you out on jobs you are wrong for or with which you have a conflict

sells you out for the quick buck

doesn't return your phone calls

doesn't negotiate for extras

doesn't get the calls

Types of Agencies

There are all types of agencies in all sizes with different specialties and each with a particular cachét. They command varying levels of respect, have contrasting methods of dealing with clients and buyers, and use different criteria when signing new talent. Despite the diversified nature of this end of the business, a few general characteristics can be singled out.

An agency may consist of as little as one person, where the agent handles all the incoming casting calls, represents all the talent, and wears all the hats associated with running a business. A large agency may be composed of tens of agents and many departments where each agent in effect acts as his own agency, representing his own client roster and working with the different casting directors on behalf of those clients. Medium-sized agencies may have many agents and several different departments, where the clients are represented equally by all the agents in a particular department. Each agent in that department may work with a different sector of the casting community.

Full-service agencies represent all types for all areas, whereas specialty agencies represent performers in only one area, such as commercials or legit.

In turn, a commercial agency or division may be noted for one or more areas of expertise, such as handling high caliber legit actors, glamour types, character actors, celebrities, extras, or voiceover artists.

An agency may have department store appeal, allowing casting directors to utilize "one-stop shopping" and get everything they need with one phone call, or it can be more akin to a boutique, offering a clientele with a specific attraction, such as interesting or cutting edge types. Some agencies are considered to be bargain basements, where casting directors can get actors cheaply and with little fuss, while others demand and sometimes get higher prices for their clients.

Finding the Right Agency

How, with so many factors to contend with, is an actor to find the agency that's right for her? Firstly, and most importantly, deal only with union franchised agencies. Secondly, research who represents whom by obtaining agency client lists from the unions. If you are not yet a member, find someone who is, and have that person get the lists. To find out who the names on the lists are, go to the *Player's Guide* or *Academy Players Directory* office (see page 51 for addresses), and leaf through their directories. This will help you get a feel for the type and extent of an agency's clientele.

An agency's client list is often a closely guarded secret, given out only to buyers and industry personnel. Commercial agents are able to handle a much greater volume of talent than their legit counterparts. If the true extent of their representation, or lack thereof, were to get out among actors, it could serve to intimidate or turn off potential clients. Because an actor is on an agency's list doesn't mean he is active commercially. He may be unavailable due to extensive conflicts, film, TV, or theatre commitments, personal reasons, or may simply be out of town. For this reason, and because of the greater volume of work and constant need for new faces in this area, a commercial agent is able to represent many people of the same type with no conflict of interest. Once you have obtained the clients lists, consult *Ross Reports* or other industry data to check the agency's ratio of clients to agents.

More important than the number of people an agency represents, however, is the *type* it represents. If its clients are all established stars and you are just beginning, this may not be the place for you.

An actor must honestly assess what stage of his career he is in. If he is just starting out, for a time any agency will do. As he gains experience and exposure he should become more selective. It is imperative that the actor stay informed through networking to learn which agents are well regarded and which ones are not up to par.

Big-name agencies such as William Morris, CAA and ICM may be right for some, but they may not be the best to build a beginner's career. It is very easy to get overlooked in an agency of that size and nature if you are not already a star. Medium-size agencies can offer you more personalized service and have the pull with the casting directors that you need, and yet you can still be overlooked. Small agencies may offer you the most in the way of personalized attention, but they may not work with all the casting directors, get the first call on a job, or have the pull to get you an appointment with someone you don't know.

As a general rule of thumb, you should work with agencies that are in the same phase of their careers as you are. If you are just starting out, you may do better with the smaller, less established agencies, and so on.

In any case, trust your instinct. If an agency feels right, then chances are it is. Keep in mind that an agent or agency that believes and supports you in your work is worth all the big name agencies in the world.

Signing *vs.* Freelancing

Signing with an agency involves putting pen to paper by two parties who hopefully are committed to making the relationship work. Freelancing involves no actual commitment, but rather is a dating process, in which the agent sends out the talent in a frequent, casual, or haphazard way. A handshake or verbal agreement may be an engagement, as a prelude to signing, or akin to living together, where both parties agree to work exclusively with one another but are not willing or able to take the final and legal step.

Some agencies are strictly signed-client only, some will freelance with you indefinitely, and some will freelance with the intention of eventually forming an exclusive arrangement. In Los Angeles the choice is limited in that actors are seen only by agency submission, and the agencies work only on a signed-client basis. In New York and other cities a myriad of options await, which can be quite confusing to those who don't know their way.

In general, it is usually better to freelance at the onset of your career before settling down with one particular agent or agency. I say "usually" because occasionally an actor is such a hot commodity that signing with the correct agent right in the beginning will not limit him, but instead keep him from making choices that may not be in his best interest and direct him toward long-term goals. Some seasoned commercial actors remain freelancers by choice, maintaining that it gives them better control over their careers, with more freedom of choice and mobility. Some actors remain freelancers because no one will sign them, due to either a bad reputation or limited salability. In some instances, it may help to freelance during stagnant stages of a career to help reestablish bearings and focus.

Unless the person doing the freelancing is a seasoned pro who knows the business and is known *by* the business, freelancing will limit the amount and caliber of commercials an actor auditions for. The freelancer may be submitted on a single job by many different agents. Technically, whoever cleared the actor, or called him to make sure they could submit his name, is supposed to get the submission. But most casting directors give the appointment to whom-ever made the suggestion first. A common scenario is as follows: The casting director and agent discuss the names on the agent's list and the agent brings up the name of a particular freelancer, let's call her Rita Smith. The casting director says, "Nice idea, but I got her from someone else first." The agent passes over Rita Smith and goes on to the next person on her list. Because Rita Smith is not "pushed" by the agent who submitted her first, either because he didn't know he was first with the submission, or he is pushing his signed clients first, Rita Smith fails to get an appointment. Not only does she not get to audition, but she never even knows she was submitted.

An agent is reluctant to push freelance talent with a casting director be-cause not only are her signed clients her first priority, but too often after convincing a buyer to see someone, the casting director realizes she got the name earlier from another source and honors that submission. Once when I was actively agenting I negotiated the terms of an agreement for a freelance actor who was the client's first choice and, according to my records, was sent in by our agency on this particular job. After many hours were spent hammer-ing out arrangements through long-distance phone calls (on our bill!) and faxes, I learned that the individual went in on the first call (there were many calls, for this was a major job) through another source. At no time through the course of the negotiation was my agency informed that this actress actually went in on the job through another agent. The terms of the deal were quite fantastic, and the other agent was delighted to collect her 10%. Part of the responsibility did of course lie with our office, but after this happens a few times in an agent's career, working with freelancers is approached cautiously and infrequently.

Freelancers have to know how to fend for themselves. Often the agents they are working with do not look out for them the way they would a signed client. The best jobs are reserved for clients, while the borderline and down-right undesirable jobs are given to freelancers. For instance, an agent might receive a breakdown for a *scale-and-a-half* spot (the union minimum for multiple conflicts on a single job) that is holding several different products as conflicts. Dairy boards are notorious for holding all beverages as conflicts, and can hold an actor's exclusivity for milk, soda, beer, juice, coffee, and tea. This is an unwieldy conflict list for a small job that won't net the actor very much money. The agent, not wanting signed clients to be restricted from doing so many commercials, submits freelancers instead. When the casting director calls back a few days later because the clients haven't found what they wanted and are ready to go *overscale*, the agent might then submit signed clients. Or, right from the outset the casting director might have said, "Submit a scale list *and* an overscale list." In such a case, the agent submits freelancers at scale and signed clients for overscale. She couldn't possibly put the freelance talent in for overscale when some other agent may be submitting the same actors at scale.

Ours is a cutthroat business, and as a freelancer you really have to have your wits about you and know what you are doing. You may decide to hold out for the better auditions, and find yourself with nothing. On the other hand, newcomers should be happy to go out on whatever they can, which is why going out through several different sources is fine in the beginning. It also gives a great overview into the workings of the different agencies.

When To Sign

Signing can be a coming together of two forces, which when combined will work together towards a common goal. An agent cannot commit to a freelancer the way he will to a signed client. When signing a new client, the scrupulous agent knows that he is partially responsible for putting food on the actor's table, and does not do so idly.

When you find an agency that has your best interests at heart, sees eye-to-eye with you on the direction of your career, and with whom you share a sense of trust and compatibility, it is probably time to think about settling down.

If you have been freelancing for a while and the calls are coming in from several different sources, it's time at least to narrow down the number of agents you are working with. If you are getting too many auditions, you have a few areas of conflict, you are being sent on the wrong jobs, agents are fighting over who submitted you first, and there are jobs around town that you are right for but are not getting in on, it is time to sign. Depending on which phase of your career you are in, working with one agent can help you prioritize and be more selective, or it can help you widen your base of operations and get in on more jobs.

Signing across the Board

Some actors feel they should sign with one agency to represent all facets of their careers. While this may be an optimum situation, it is not always possible, necessary, nor in your best interests. Most agencies have different areas of strength. Some agencies may be powerhouses in the legit arena, but have inconsequential commercial departments. Agencies known for strong commercial departments may hold no weight in the legit market. Some agencies may have strong overscale or celebrity divisions, but do not compete on the scale commercial jobs. Not only might an agency have different strengths, but an actor is rarely at the same stage in all areas of his career at once. So, too, an actor may love the legit agents in one office, but fail to connect with that agency's commercial agents. An agency may feel strongly about representing an actor for commercials, but may be unable to represent him for legit because he is either too green or conflicts with someone already on their legit client list.

For the best and most individualized representation, an actor should consider signing with one agency for commercials and another for legit. If both agencies are competent and have an actor's best interests at heart, they should have no trouble consulting with one another to iron out scheduling difficulties or other work-related conflicts.

Sometimes a viable commercial actor encounters agents who will sign him for legit only if the actor places his commercial business with them as well. This is a tactic used by agencies to shore up weak commercial divisions. The truth of the matter is, however, that if an agency feels strongly about an actor's legit salability they should and will send him out no matter who represents him commercially. If they only want him for his commercial salability, and are offering legit representation as an enticement to sign, they may not wholeheartedly represent him for legit, and that actor may be much better off taking his business elsewhere. An actor might argue that this is the only way he can get legit representation. While it may seem a viable solution, rarely does this type of arrangement benefit the actor. Instead he finds his commercial business whittled down, and after an initial few legit auditions, may find that area closed off to him just as it was before.

The Hispanic and Minority Markets

Some actors may be marketable in the Hispanic market and reluctant to sign with mainstream agencies who don't get a large portion of work in this area. If this is your situation, level with your prospective agent. She may be willing to let you continue to work in this area through other sources while holding you to exclusivity on all non-Hispanic television work.

Other actors are offered representation exclusively for minority work. While this is one way to go, realize that the more mainstream options will not be available to you since this type of agency strictly gets minority calls.

Industrials

Many commercial agents do quite a bit of industrial work, either taped or live or both. Taped industrials, which are shown in the workplace, are done through AFTRA, and live, which are attended by industry personnel, are done through Equity. Taped industrials generally use spokespeople and a wide variety of office types. Live industrials typically use spokespeople and singer/dancers. (See Chapter 9 on industrials.) Commercial agents may or may not ask you to sign for these areas while still submitting you on jobs. For many actors this will not be an issue. Rather they are pleased to have the bonus of being submitted on industrials as well as commercials. Some performers, however, have developed thriving industrial careers on their own. They get called directly or through agents who do a high volume of work in this area and are not necessarily the same agents sending them out on commercials. They may not want agency representation in this area, although a good agent can help the practiced actor command more money and better terms than he is able to negotiate on his own. If you do not want to be held to exclusivity for industrials, talk to your agent. She may have no problem allowing you to freelance in this area.

Some actors may be going out on Equity industrials through their legit agents. These agents may not have a problem with the actor going out in this area through his commercial agent as well. Others may not want another agent infringing on what they feel is their territory. Still others may not want anything to do with industrials. In any situation the name of the game is communication. Talk to all concerned and find out how everyone feels on this issue.

Handshake Agreements

Some actors are reluctant to sign contracts and may request a handshake agreement until a permanent agreement can be reached. Handshakes, generally used by actors or agents not ready to make the final commitment, may work for a short period of time; but if the arrangement is prolonged, the energy may go out of the relationship before it ever actually begins.

Instead, it is probably better to jump in when the timing and the agency are right. Capitalize on the heat that is generated when two parties are excited about taking the plunge together, instead of seeing it wane because one party cannot make up his mind.

Once again, communicate. If both parties genuinely want to make an arrangement work, a little give and take will make any situation palatable to all concerned.

Agency Contracts

Franchised commercial agents (and these are the only kind you should be dealing with) will ask you to sign standard SAG and AFTRA agency contracts. There is no deviation allowed on these, so it is not necessary to take them to a lawyer. However, you should be familiar with them in order to understand your options.

Agents are allowed to take 10% of your gross earnings.

The first time you sign with a particular agency, the commitment is for one year. Subsequent renewals with the same agency can be for one to three years. Most agents will ask that you renew for three years, but if there is some question about the relationship, one or two is an option.

Regardless of how long you are signing for, the unions provide you with an *out clause*. If an actor does not make a minimum of $3500 in any 91-day period, or receive an offer of work in the first five months, the actor or agency can dissolve the contract. The following is an example of such a clause.

> *If, during the period of 91 days immediately preceeding the giving of notice of termination, the actor fails to receive compensation in the sum of $3,500 or more for services and reuse fees for commercials in which the actor was employed during the term of this contract then the actor or the agent may terminate the engagement of the agent by written notice to the other party; provided, however, that if this contract is not a renewal contract, then such notice may not be given prior to 151 days after the date of commencement of the term of this agency contract; provided further that if no bona fide offer of employment for television commercials is received by the actor within any consecutive period of 120 days during the initial 151 day period then either the actor or the agent may terminate the engagement of the agent hereunder by written notice to the other party.*

In addition to the out clause, any worthwhile agent should not hold an unhappy actor to contract. It is unfair and not worth the bad press.

Personal Managers

According to the National Conference of Personal Managers (NCOPM), "The personal manager is the person who helps the artist in his or her creative endeavors, who nurtures, grooms, guides, befriends, and at all times is the driving force in the forefront, breaking through the barriers of frustration and difficulty that one so often encounters in the entertainment industry."

On the down side, personal managers are not franchised or regulated by the unions, nor do they have to be licensed by the state. Because of this, they are able to charge higher commissions and can therefore command a smaller client base than agents, who can collect no more than 10% and thus need more clients to produce more income. The manager's primary function is to give personalized attention and direction to an actor's career. This attention is sometimes limited to advice-giving and hand-holding as the personal manager is prohibited by law from soliciting work on his client's behalf and is not allowed to negotiate directly with the principals involved if there is an agent of record or if the actor is signed to an agency. Managers can and do charge upwards of 25%, and, unbelievably, a few charge even more. While NCOPM is an association dedicated to organizing, holding in check, and offering support to its personal managing membership, it is still somewhat of a fledgling organization, and many managers do not yet belong.

With the possible exception of child actors, many people believe that personal managers are simply not necessary at the onset of a career. (Children's managers are discussed in Chapter 7.) While getting an introduction to an agent is a service provided by a personal manager, this is not anything you cannot do on your own. Unless you are established, having a manager is not always well regarded by agents and may even serve as a deterrent to your being signed.

A personal manager can be invaluable if the actor has reached a point in his career where he feels he needs more guidance and individual attention than he is getting from an agent. At this stage a manager can also help keep the actor's business affairs in order.

Stories abound about unsavory managers embezzling or losing actors' money. In addition, some, even the more reputable managers, have actors sign what amount to unconscionable contracts. A common tactic is to have actors sign long-term contracts, possibly five years, to be renewed only at the manager's discretion. This means that when the term of the contract has elapsed, the manager can legally keep the actor on as a client whether that actor wants to renew or not. Breaking these contracts is costly, and often the actor's only option is to buy off the manager with a sum of money.

At the right time and under the right circumstances, a personal manager can work wonders for a career. If you are interested in pursuing this avenue, you must follow a few guidelines:

1. Do not *ever* pay money up front; a fee is taken when money earned comes in.
2. Do not have pictures done with a photographer or take classes with someone as a condition of becoming a client.
3. Have the contract reviewed by a knowledgeable entertainment lawyer.
4. Check to see if the manager is a member of the National Conference of Personal Managers, 1650 Broadway, New York, NY 10019, or 4527 Park Allegra, Calabese Park, CA 91302. (Many reputable, established managers are not members of this organization, but if the manager you are checking on is a member, you at least have a place to start. See the Appendix for NCOPM's code of ethics.)
5. Make sure that the manager is well regarded and reputable. Check with agents, the Better Business Bureau, the State Attorney's office and other actors.
6. Get a client list before signing, and see if you can track down and talk to some of these people as well. Remember, however, that the manager isn't likely to release the name of anyone he is having legal troubles with.

The Agency Interview

The agency interview is the moment you have been waiting for—your chance to impress a potential agent with your wit, charm, and talent. It is understandable that an actor be nervous before a meeting of such import, but nervousness is a distraction that masks the real you.

Coping with Nervousness

Instead of giving your nervousness free rein, try channeling this energy to work for you. With practice, you can use it to give you additional spirit, élan, intensity, or power. Do preparation to keep calm before an interview. Meditate at home or in the bathroom at the agency. This involves sitting calmly and trying to clear your mind of all but one point of focus. Try concentrating on the image of a candle flame. With time, you should be able to keep your mind from leaping around and eventually connect with your innermost thoughts.

You can work on yoga breathing exercises such as the alternate breath, which is simple and can be done in a waiting room full of people without attracting too much attention. Close off one nostril with your thumb. Inhale deeply through the other nostril to the count of eight. Hold for the count of eight. Exhale for the count of eight. Repeat on the other side. Doing this for a few minutes will clear your thoughts before meditation and will help you find your center and concentrate on the task at hand.

Try positive imagery—imagining the situation as you would like it to go—or affirmations—repeating to yourself that you are wonderful and talented and that the interview is going to go fantastically well. Many actors swear by the Bach Flower Rescue Remedy, a few drops of which they put under the tongue before emotionally tense situations. It is an inexpensive, completely natural flower extraction, which apparently works on an emotional level. It can be found at health food stores. People who have used it before television appearances, speaking engagements, and even during childbirth, have found it to be very effective for calming the nerves.

What an Agent Wants To See and Hear from You

Realize that the agent wants you to do well. She wouldn't be holding an interview unless she hoped to gain something from it. What she wants to see is a sparkling and forceful personality, a developed sense of self and confidence, and a basic understanding of the business and your place in it. She would like to see an honest assessment of your talent and appeal and an objectivity about who you are. She is attracted to honesty, openness, humor, and warmth.

What an Agent Does Not Want To See and Hear from You

What she does not want to see is pessimism, defeatism, or lack of energy. She does not want to hear you insist you are one type when everyone (including her) sees you as something else. She does not want to shake a dead fish when she offers you her hand. She does not want to be overcome by your perfume, aftershave, or body odor. She does not want to have to explain the business from the ground up. She does not want to dig your picture and resume out of the file because you assumed she had it on hand.

Ten Steps to a Great Interview

1. Be prepared. Find out anything and everything you can about the agency and individual you are going to see. Ask other actors about their interviews at that particular agency. Ask casting directors about them. Go to a drama bookstore and browse through their information on agents.

Ask the clerk what he knows. The clerks at these bookstores tend to be fountains of information. If you haven't already done so, get the agency client list from SAG and look through the *Player's Guide* or *Academy Directory* to see who they represent.

2. Dress as close to your type as possible without overdoing it or masking the real you. If you are a blue-collar type you don't have to come in wearing work boots and a hard hat. A simple open-necked shirt will do.

3 . Bring several pictures and resumes. Do not expect the one you sent in three months ago to be there waiting for you. The extras are necessary in the event that the agent refers you on to see other people in the office during this visit. You may be right for a job she is currently working on and lucky enough to be sent right out to audition, in which case you will also need an extra picture. Bring your videotape if you have one, in case the agent wants to see it. Be prepared to leave it, and make sure it isn't an original. Label it with your name and phone number before meeting with the agent.

4. Be prepared to wait. Busy offices have been known to keep actors waiting an hour or more. Do not be insulted if this happens to you. They are merely attending to priorities, and meeting new talent happens to be at the bottom of the list. Bring a book. Do not go out for coffee or the agent may call you while you are out. Do not spend too long a time in the bathroom or they may think you left. Find a way to keep up your energy and confidence up during this waiting period.

5. Bring a list of current conflicts if you have any, and a list of casting directors who know your work. You might also jot down any auditions or callbacks you have received lately. Do not trust your memory.

6. Be ready to talk about yourself. This does not mean to relay your entire life history, merely to talk about related events. Start with where you went to school, how long you've been in town, which shows you've done, which agents you've been in to see. The agent really just wants to hear you speak and get to know you a little.

7. Try to find areas of common interest. Not only does this make for livelier discussion, but it will help the agent remember you.

8. Keep focused and keep your goals in mind. This does not mean you should say, "What I'd really like to do is film...Could you set me up with your legit agents?" It's a given that everyone wants to do film. Your goal for this interview is to impress the agent with your commercial marketability and potential.

9. Be yourself. Meet the interviewer's eye. Offer a firm handshake. Show eagerness. Be polite and enthusiastic. Don't oversell. Ask questions, but don't interview the interviewer. Don't chew gum. Don't smoke, even if invited.

10. Ask how the agent wants you to follow up. Ask if she would like to see your videotape. If the interview has gone well, you might ask if there is anyone else in the office she would like to meet.

Follow-Up

Follow up your interview with a thank-you note right away. Only if the agent has been especially kind or helpful to you is even a very small gift appropriate. Most agents are wary of gifts of any kind. While by no means essential, as a kind and appropriate measure, the proper memento can give the agent one more way with which to remember you.

During an interview with a young actor by the name of Taylor Nichols, who I had asked in after being impressed by his work in a commercial acting class, we discovered a favorite author in common. A few days after the meeting a paperback copy of Cider House Rules *by John Irving found its way onto my desk. I found this to be a thoughtful and unassuming gesture. Another actor used to leave cartons of orange juice, with funny notes attached, on my desk. It brightened my day and gave me something to drink! While neither of these actions caused me to send out the actor more than his due, they may have helped me to think of them more often, or perhaps in a slightly better light.*

Two weeks after the interview, follow up with a postcard. Call only if the agent said it was okay to check in.

Assuming that all went well with the interview, the agent will probably want to freelance with you, either at length or as a precursor to signing. It is up to you to evaluate the situation and determine which scenario is in your best interests. As discussed before, if it feels right, and the agency and timing seem appropriate for your needs, than chances are it is.

After Signing

Once you have signed with an agency your work has only just begun. You now have someone else in your corner, but it is up to you to help the relationship run smoothly. After signing, make sure you know everyone in the department and that they know you. You should by now be known to all the commercial agents, or meetings should be set up immediately. Ask to be taken around to meet the assistants. Make note of their names and which agents they work for. Ask to meet the business manager and president of the company if you don't already know them. You may not have any dealings with the president, so a meeting is not crucial at this point. You will eventually be dealing with the business manager, so it is a good idea to be introduced.

If you are signing with a full-service agency for commercials only and still need representation for legit, ask if a meeting with someone from the legit department might be set up. If the idea is not taken up gracefully, do not press the issue at this point. They are in no way obligated to find you legit representation, and there may be something going on between departments that you don't know about. Or you may not be strong enough in that area and your commercial agents may want you to wait until you can make a better impression. Don't be afraid to pursue this later, once you and your agents are on sure footing.

An agent collects 10% of everything you bring in, so they should not be expected to do three quarters or even one half of the work. Keep drumming up leads, and refer them to your agent. Send postcards to casting directors and producers about your new representation. Make sure your agents know which casting directors do not know you or your work, and ask if they can set up general interviews. This is not that common a practice in the commercial area, but it can't hurt to check.

Your agent will need a list of any current commercials, identification numbers, and the ad agencies they are through. This is so that they can keep track of payments and notify them when it is time for renegotiation. If an ad agency is not notified of your intent to renegotiate within 90 days of the date the commercial is due to elapse, they are allowed to renew it at scale. This is not in your best interests, as renegotiated commercials often result in the actor making a substantial amount of money.

Doing Your Part

Depending on what type you are, how active the business is, and a myriad of other variables beyond your control, you may hear from the agent immediately or not for several days. Resist the urge to call and ask, "What's up?" Check your machine and service hourly, get call waiting on your phone. Keep yourself busy by stirring up contacts and taking classes. Let the agent see you doing your part. Keep honing your craft so that when the calls come in you will be ready. Take a commercial class if the agent recommends it. Take acting technique, speech, dance, or vocal instruction. Work on your skills. Practice roller skating and teach yourself to juggle. In short: Do not sit by the phone and wait for it to ring!

Keep your agents apprised of any strides you are making in your legit career. If a particular legit casting director or producer knows and likes your work, tell your commercial agent. If you have come close on any jobs, or have been seen for anything with any weight, your commercial agents will be interested. If you are in any shows, whether in or out of town, your agent should know. Provide them with copies of your reviews. Every piece of information will give them more ammunition with which to sell you.

Let your agents know ahead of time when you will be unavailable or out of town, if even for a day. If it is for any length of time, call again the day before and remind them. Give them numbers with which to contact you.

Be ready to go on auditions at a moment's notice. Casting directors often call with last minute changes or requests, or your agent may be using you as a substitute for another actor's cancellation in order to get you in to see a casting director who doesn't yet know you.

If you are running late to an audition, get to a phone and let your agent know. She will notify the casting director for you. Otherwise, you may not be seen when you get there. If you can't make an audition, let them know as far in advance as possible so that they can fill the spot with a replacement.

Keeping Visible and in Touch with Your Agents

It is also up to you to keep your face fresh and in front of your agents as much as possible. By being inventive and considerate you can do this in a positive and endearing way. Drop by their office to show off a new haircut or manner of dressing:; stop in with new photos or to check on your file. Make sure it is well stocked and that the resumes are updated and attached and trimmed to your pictures. Even if you are on very good footing with your agents, try not to go in simply to say "Hi," but rather for a reason. If you can't think of anything else go in with some homemade cookies or fresh fruit. The hectic pace of the business leaves little time for eating and assistants especially are usually ravenous. Remember too that assistants are the agents of tomorrow. They may also be helpful in rescheduling appointments for you or providing any number of professional courtesies. It can't hurt to stay on their good side. Since they are usually overworked and underpaid, gifts of food are simply a nice gesture on your part.

> When I first started out in the agency business, I worked for two of the office's busiest agents. I used to put in 11-hour days and rarely took time out for the bathroom, much less for food. Before the more updated phone systems of today we had phones with buttons that lit up for each call. Often there were ten or more calls holding for a single agent, and we would just go from one to another without taking a breath of air. Once an actor by the name of Barry Kivel saw me through the glass partition and must have taken pity on me. He disappeared for a few minutes and returned with an enormous bran muffin, which he silently left on my desk. I have never forgotten this thoughtful gesture and have always been delighted to hear from Barry.

The most simple act can leave the most vivid impression. This can work against you as well. If you are pushy, overly demanding, or rude, your agents will not be so pleased to represent you. They may still do their job as your representatives, but they will probably not go to the nth degree for you.

If you feel out of touch with your agent(s), you can suggest a meeting, or if you feel so inclined, try inviting them to lunch—either singly or as a group. A neighborhood coffee shop will do. One actress from our office traditionally provided us with a post-Christmas breakfast of bagels and lox. Not only does this arrangement provide a pleasant scenario in which to relax and chat with your agents, but it helps them to get to know you on a more personal level.

Holidays in general, are a big getting-in-touch time. They provide the perfect occasion to touch base with your agents as the work slows down at holiday time. Drop by an agency office at Christmas time and chances are that you will see it stocked with all sorts of delicious and fattening goodies. If you want to do something of this sort, try doing it well before or after the actual holiday. Otherwise your efforts may not get the proper notice or appreciation, as most agency people are pretty well gorged by Christmas time.

You shouldn't have to bribe your agents with gifts or ply them with food to keep them working and in your corner. Expensive or inappropriate gifts are not called for, and may rather serve to put off your agents and make them uncomfortable. As mentioned before, a small, inexpensive, or homemade token of your goodwill or appreciation is a nice gesture and will serve you well. On the other hand, if the agent bends over backwards and gets you a deal beyond your wildest expectations, it can't hurt to send flowers or some other expression of your gratitude. You may say, "Well, that's their job," but agents don't often personally see the results of their negotiations. It is good business sense to reward a job well done. This is of course completely up to you.

A way of keeping yourself from being overlooked is to become visible in other areas of the industry. An active role will give you more topics of discussion, more reasons to be in touch with industry people, and will keep you visible in a roundabout way.

All in all, if you are constantly active in improving yourself and your skills, are visible within the industry, do your best on auditions and callbacks, continue networking and making contacts, communicate with your agent and let her know when things are bothering you, show up on time and return calls promptly, act professional and courteous in dealing with people, then you are doing your part. But how do you know when an agent's not pulling her share of the load?

When To Worry

The agent with whom you first met, initiated your signing, or with whom you feel most comfortable is your primary contact within the agency. If any problems should arise with regard to your representation, it should be brought to this agent's attention first. If all but one agent in the department call you for auditions, then let this be known. The primary agent can feel out the situation for you and determine if there is a problem. Perhaps the agent who is not sending you out doesn't know you that well, has overlooked you, or is not confident in your abilities. *All* agents within the department (if that is indeed how the agency operates) should be submitting you for jobs.

Sometimes, however, there are reasons for going out more through one agent than others. There may be a hierarchy of sorts within an agency, where one agent works with the most powerful casting directors, the next agent works with the next level, and so on. Depending on your professional level, you may find yourself going out through one agent more than the others. In some cases each agent may concentrate in a different area, where one works primarily on beauty spots, another's casting directors may do primarily P&G work, and another may do mostly comedy or character work. For this reason, also, you may find a preponderance of auditions coming through primarily one or two agents within an agency. Whatever the situation, it is up to your primary agent to identify the problem, if there is one, and smooth over any difficulties.

If the agent or agents seem to be lacking in imagination or vision, this should also be worked out by bringing your feelings to the forefront. If casting directors are saying "Why didn't I see you on this...?," or "Why don't you ever go up on this type of role...," if there are roles out there you *positively* know you are right for, or if you know you can go out on more types of parts than what you are being sent out on, speak to your agent. You may be right, and they may try to broaden their submissions for you. They may greatly disagree, in which case they should give you their reasons why, or the casting directors may simply be refusing to see you in any other way, and then it's up to your agents to help change the situation.

If the agents or assistants give out erroneous or incomplete booking information, this too should be brought to the agent's attention. It may not be your agent's fault; the casting director may have given out the incorrect information. It could be that the assistant is not up to the task of delivering information properly and the agent will have to prepare you directly in the future. If it continues, the agency may simply be slipshod in its methods, which puts you at a severe disadvantage competitively.

If you are constantly up on jobs where the casting director asks, "Why are *you* here?" or where everyone else on the audition is very different from you, the agent either hasn't seen you in awhile, is confusing you with someone else, is mistaken in his conception of you, or is simply incompetent. Mistakes do happen, like the time I sent a blonde-haired, blue-eyed actor on a spokesman call, only for him to call back and politely inquire why he was sent out on that particular audition. "What do you mean?" I asked. "Well," came the embarrassed response, "I was the only white person there." It turns out that the specs on the job had changed at the last minute and the casting director had forgotten to notify our office. Oops.

Many of these situations do arise from time to time. But if they arise with consistency, and are still occurring after you've brought it to your agent's attention, perhaps it is time to think about switching agencies.

Interview with Vincent Cirrincione
Personal Manager
Cirrincione Lee Entertainment
300 W. 55th St., New York, NY 10019
200 N. Robertson, Beverly Hills, CA 90211

BEARDSLEY: How did you get into the personal managing field?

CIRRINCIONE: I was in the music business for a lot of years managing musicians. I just got tired of it. The music business was really crazy, even more crazy than it is now. I always wanted to be an actor. It was just by osmosis that I started doing this, and I opened up an office.

BEARDSLEY: What is the function of the personal manager?

CIRRINCIONE: I'm the conduit, the buffer. I make sure that my clients get the utmost of opportunities, and I advise them.

BEARDSLEY: What do you say to people who claim that you do not need a personal manager?

CIRRINCIONE: In any business you need people who are going to help you, and the more people you have to advise you the better. An agent may help get a particular client a job, but they may also get opportunities that I might not think are right for a career. We turn down jobs worth a lot of money because they are not right at a particular point in a career. You know what it's like working in commercials. The idea is just to get the job. I even have a letter from one agency telling people to take test shots because business is so bad.

No manager or agent gets the job for a client. The *client* gets the job. *We* try to help direct and build their careers—to advise them. Their talent makes their career. But you definitely can be very responsible in helping them choose the right projects and keeping them in the state of mind that they are *out there*. When people say you don't need a manager, they don't know the business. It is very simple: Once you decide to be in this business you have a career. You might not be making money. In any career you need to be as much as you can be. Managers like us are very different from agents and managers of ballplayers, because they have nothing to do with the talent. They don't do anything like advise which college to go to; they have nothing to do with the player's development.

BEARDSLEY: What do you do when the business is bad and your client needs money?

CIRRINCIONE: I try not to base my decisions on money, although ultimately everything *will* be money. We all want to make money. But sometimes you don't go for the short return, you go for the long money. I was just offered a large sum for four weeks work for a client. The part is not right for her, in fact it is terrible. I had to say no, because I am trying to build her career.

BEARDSLEY: How many clients do you represent?

CIRRINCIONE: It varies, but right now I have maybe seven or eight main clients on the East Coast and about the same on the West Coast. But when I say main clients I am not talking about the people I am trying to develop.

BEARDSLEY: How long would you develop someone before realizing it might not work out?

CIRRINCIONE: I have had one client with me for nine years, who has always worked, but is just now making a lot of money. Now he is in a Broadway show, he's on a soap opera, and he does voiceover work for commercials.

BEARDSLEY: How do you find talent?

CIRRINCIONE: In the case of Halle Berry, I went looking for her. Years ago I thought there was going to be a big upswing in black movies. I told a friend of mine that I wanted to find an actress or actor who was young and black. Halle was a pageant girl. I called the director of the pageant, who then put me in touch with her.

BEARDSLEY: What do actors or actresses do wrong?

CIRRINCIONE: There are a lot of people who walk through the door and are not prepared. When someone says, "I can't afford it," or "I can't take lessons, I have to work," they are not going to be prepared. A baseball player doesn't seal it the first day. He goes to spring training. If he doesn't do well in spring training, he doesn't have a good season. Actors always have to be in the game.

BEARDSLEY: What gets you excited in a new talent? What do you look for?

CIRRINCIONE: I think there is a spark. You meet a lot of people, but they all can't be superstars. When someone walks in with a spark and lights up the room— it's just there. With Halle, I saw something, special qualities that she had, and now everybody is breaking down doors to get to her.

BEARDSLEY: Still there are some agents who say that the buffer, as you called it, of having a personal manager removes them from direct contact with the client.

CIRRINCIONE: You know who says that? The insecure agents. It is just not true. Any good agent will realize that you keep contact. A lot of times an agent doesn't want someone looking over her shoulder. But do you know what? Agents don't pay me, the client does. Maybe the talent wants someone to look over the agent's shoulder. Certain agencies just put people on soaps because that's what they do. They promise this and that, but it doesn't happen. I'd say, "Listen, you have been on soaps long enough, let's leave." I've turned down jobs for clients where, if my motivation was to make money right then and there, I would say take it or think it over. I'm in business, I want to make money, I'm not saying that I don't, but a lot of agents go for the short money. There are a lot of good agents, but for the most part their nature is to get work, not build careers.

BEARDSLEY: What do you think about the problem of managers charging sky-high fees?

CIRRINCIONE: I do not charge a sky-high fee. I work for my money; it is not a gift. It's a business. Let me put it into very basic form: If you have a store on Fifth Avenue, you are going to pay a lot more rent than if it's on Ninth. You don't say, "Well, gee, how come I have to pay so much rent?" Because I am Tiffany on Fifth Avenue. If I am Tiffany on Ninth Avenue I'm not going to do any business, and I'm not going to be able to sell the product.

BEARDSLEY: You charge 15%, which is not that high compared with some manager's fees, yet on top of an agent's 10%, taxes, and so on, the actor is left with only a small percentage.

CIRRINCIONE: If you run a business, let's say a grocery store, you don't work on 80% or 90%, you wind up working for 18% or 20%, even 12%. They don't work on 30–40%. Clients have to pay agents, lawyers, managers, and they still come home with 35–40%. That's a large percentage of your business. You have to think of yourself as a business.

BEARDSLEY: What is a fair contract in terms of length of time?

CIRRINCIONE: A manager's contract usually is three to five years. I have had both scenarios—where for three years nothing happened for a client, and where things have happened right off the bat. Rubin Santiago Hudson has been with me nine years, and although he was always a working actor, he is just now starting to make a better-than-average living. I believed in him.

Now Halle is the other side of the coin. She walked in and got a job. Every year she kept making more money and will be into seven figures next year. That's why five, six years is not that long when building a career. Basically, if I do my job as a manager, I turn down money. I have an offer on my desk for $100,000 a week, and I turned it down because it stinks. I'm turning down $32,000 for me because it is not the right thing for her career. To turn down money daily you have to feel secure for the long term. If it's not secure, and it's only going to be for a short time, then you might take the money and run. It's a matter of commitment.

BEARDSLEY: What about contracts that are "self-renewing?" In other words, contracts that managers can renew solely at their own discretion?

CIRRINCIONE: Everything in this business is in favor of the person doing the buying. That is basically the bottom line. If you don't want to be managed by someone, you make a deal to get out of the contract. Basically you shouldn't be involved with anyone you don't really trust. Unfortunately in this business, you have to rely on your gut feelings until you really get to know someone. My contracts are for three years. When they are up they are subject to renewal of another three years at the client's option. Most of the money doesn't come until further along in a client's career. Halle Berry has done everything you would want in a storybook way, but the big financial payoff is still in her future. If she hadn't shown loyalty to me, all the work would have been in vain. But I still would have done it, because that's my job.

BEARDSLEY: Do you feel that working in or having worked in commercials is detrimental to an actor's career?

CIRRINCIONE: I think commercials are helpful. You might do commercials for money, but it also puts you in front of the camera. Kevin Costner was called in to read for his first film after being seen on an Apple Computer commercial. This business is like any business in that it requires a lot of luck, and you have to be ready for that luck. There is an agent here who just moved from one agency to another. She said that she didn't even realize she was in the same business—that she has *access* now. Access means that she knows about things before they turn up in Breakdown [casting notices distributed to agents]. That means that if she has someone, she can start on the ground floor with him when something is first going on. Maybe the director can meet her client a year before he meets anyone else. As you gain access, you know more of what's going on. In any business, knowledge is power. Of course, if you are just starting out, you take anything you can.

BEARDSLEY: Does an actor have to be in L.A. to "make it?"

CIRRINCIONE: Saying you can't make it in New York City is not true. Halle got a television series here in New York. She did not have to move to L.A. All you have to go to L.A. for is if you haven't got a tape and are getting older, you can get a lot of work if you want to be in TV. If you don't want to be in TV then you don't have to go to L.A. It's a very simple situation. There is no one way of getting some place. Everybody gets there differently, but there is a common way of staying there. I just signed one up-and-coming client with a big agency not because the other agencies are bad but because, among other things, this agency had a lot of well known female clients. Because the stars can't do everything, we have access to scripts they turn down. Humphrey Bogart did things because George Raft did not want to do them. So you can always get things people don't want. If you want to place bets, Julia and Jody can only do one project a year. Meanwhile, maybe three other projects are great, and my client gets one of those. Then it is reversed and all of sudden my client is turning down work. Do you know that Julia Roberts was not the first choice for *Pretty Woman*? Nobody signed Julia Roberts in New York. Julia got *Mystic Pizza* when she was a freelance client at William Morris. They hip-pocketed her. Then she got *Steel Magnolias*, and all of a sudden that was it.

BEARDSLEY: What do you mean by "hip-pocketed?"

CIRRINCIONE: You have to sign with an agency in L.A., so when they don't really want to sign you, they hip-pocket you. They sort of think about you for things if anything comes along. They won't commit themselves. In New York you can freelance, so it is not a problem.

BEARDSLEY: Is there anything you want to say to actors?

CIRRINCIONE: Everyone has different needs in their lives. Some people are married, some people aren't, some have children—there is no one thing. Actors should not listen to other actors because a lot of it is based on jealousy and a lot of it is based on having different careers. No two people can have the same career. This guy can be the same age as that guy and yet the career will be totally different. In Charles Grodin's book, *It Would Be So Nice If You Weren't Here*, he says that there were people in class who had more talent, but they didn't go out and do things. Fame doesn't always go to the most talented. It's like the musician who stays in his basement all his life and goes around saying how much better he is than the Beatles, except that they are successful. It is just a matter of marketing. Some people have different personalities. My position adds another dimension to it, another person an actor can trust. My job is to keep them in the ballgame. But there are different ways of getting into the ballpark.

5

□ □ □
□ □ □
□ □ □

The Business of Show Business

Packaging your product and selling it are only parts of becoming a successful actor in commercials. Understanding the mechanics of the business, including the details others take for granted, is a vital and integral part of the process. Too many actors are ignorant of the goings-on that directly affect their careers and incomes, happy to leave these details to third parties. While there may in fact be someone competent tending to business aspects of your career, being informed will help you maintain control and keep from being taken advantage of. Knowledge is a powerful tool, adding to your overall competence as an actor, fleshing out your persona, and providing another edge over the competition.

The Unions

The actors' unions are the Screen Actors Guild (SAG), American Federation of Television and Radio Artists (AFTRA), and Actors Equity Association (Equity). SAG governs filmed movies and commercials, AFTRA, live and taped commercials, television shows, radio, and taped industrials, and Equity, live performances. Where SAG and AFTRA overlap—SAG governs filmed commercial advertisements and AFTRA holds jurisdiction over taped spots—the SAG contract acts for both.

The unions insure that their members receive favorable working conditions and terms. They also provide health insurance and pension benefits and enact guidelines in the affirmative action areas—largely responsible for the changing and growing role of minorities, women, and the handicapped in the business. The unions offer help in avoiding scams, are great forums for networking, gathering information, career guidance and meeting industry people. Many of the offices have lounges where actors gather to touch base or exchange information. The bulletin boards have information on apartments to rent, ride shares, and support groups. The union can be a lifeline for those new to the area, the business, or simply in need of familiar and comfortable surroundings. The unions hold open forums, where industry people come in to meet with new talent. SAG New York makes video equipment available to its members, and many offices offer classes and career guidance. They offer alternative career help for those looking to leave the business. They all make information available to actors, such as lists of franchised agents and their clients, union-approved producers and production companies, and warning brochures.

The three unions are run by their members through weekly or monthly meetings. Actors are encouraged to take an active role in the governing of their unions.

How to Join

The unions were set up to protect professional actors, help them along in their careers, and to accord them preference when casting for jobs. Generally, in order to join, you must either show the capacity to be hired, or already have an offer of employment. AFTRA's open door policy, however, offers membership to anyone who pays the initial fee and subsequent dues. If you have been a member of one of the other unions for at least a year and have worked once as a principal in either union, you may apply for membership at SAG. If you receive a union job offer in SAG's jurisdiction, you will also be eligible to join. A year's membership in one of the other unions and having performed at least one role comparable to Equity principal work qualifies you for Equity membership.

The fee for joining AFTRA is $600, with semiannual dues paid on a sliding scale amount based on your previous year's income, beginning at $28.50. SAG's initial fee is $796, with semiannual dues payments of $42.50. Equity's initiation fee is $500, with a standard dues payment of $52 a year, and an additional dues payment of 2% of gross Equity earnings per year.

Waivers

When you book a job on a union commercial and are not yet a member, a waiver of preference of employment must be obtained. This involves a phone call from the producer to the union with enough background information on you to justify your hiring. It is a routine and relatively simple procedure, and rarely is anyone with any acting background denied a waiver. According to the union, each case is evaluated on its own, with permission granted to qualified professionals.

Must-Join

Once you have completed your first union job, you have a month in which to work freely on union jobs before becoming a *must-join*. If you are not offered union employment within a month of the first job, you simply join whenever you get your second offer. You will not be able to work on the set of the second job until your dues are paid.

Station 12

If you are delinquent on a dues payment your prospective employer will be informed of your "station 12" status and will be prohibited from employing you until steps are taken to rectify the situation. Job offers do occur at the last minute, sometimes the evening before a shoot. If your station 12 status is discovered at the last minute, the people involved with the shoot will become frantic and upset. To prevent this from happening, pay your dues on time and keep track of your status. If you seem to be coming close to a job, call the union and make sure you are not in arrears. If you don't have the money, check to see if your agent or manager will front the dues payment for you.

The Benefits and Drawbacks of Being a Member

Besides reaping the benefits of the aforementioned hard-won advantages, being a member of the union will allow you to do extra work on union spots. What you will not be able to do as a union member is work on nonunion commercials. As this is a fertile training ground and a stepping stone into the professional commercial arena, you might want to postpone your membership until you become a must-join. Assured of some money coming in at this time, you will then be in a better position to pay the initiation fee.

Finding an Agent as a Nonunion Member

Union-franchised agents are prohibited from working on nonunion work, and so they prefer prospective clients to be union members. Considering the ease with which waivers are obtained, however, most agents have no objections to an actor waiting to join until he books his second job. Commercial agents who tell you that you must be a union member before they will send you out on jobs are either ignorant of the procedure or are looking for an easy way of telling you they don't wish to work with you.

Union Guidelines

The following information can be found in greater detail in the SAG Commercial Contract Code.

Terms of Employment

On-camera principal performer: Anyone who is seen and speaks a line or lines of dialogue, or who is seen and is identified with the product or service, or who is identifiable.

Off-camera principal performer: Anyone whose voice is used off camera except *omnies*.

Omnies: Atmospheric words or sounds uttered by anyone.

Extra performers: Persons appearing in the foreground solely as atmosphere and who do not speak, are not identified with the product or service, or are not identifiable.

Scale: The minimum payment as set forth by the union.

Exclusivity: The right of the advertiser to prohibit principal performers from accepting employment in commercials advertising any competitive product or service. This does not include products or services simply made or offered by the same advertiser or merely manufactured or offered by another advertiser competitive in some other product or service area.

Public service announcements/Government agency messages: Known as PSAs or GAMs, these are messages produced for governmental agencies such as the Armed Services and nonprofit welfare and public service agencies. Principal performers are entitled to receive a scale session payment for work on such spots, are not held to exclusivity, and do not receive residuals.

Test market: Commercials that will be used to test the product in a market. Performer is entitled to session and use fees.

Non-air: Commercials not intended for broadcast use, such as nonbroadcast audience reaction commercials, copy testing, or client demos. If exclusivity is required, the performer is entitled to standard session and holding fees. Where no exclusivity is required, the performer is paid a lesser session fee.

Session fee: Pertains to the compensation paid to the performer per 8-hour day, which also constitutes payment for the first commercial. If two commercials are made from the performer's work in one 8-hour period, he is entitled to two session fees. If he makes one commercial over the course of three days, he is entitled to three session fees. If he works for three days, and three commercials are made from his work, he collects three session fees.

Downgrading: When a principal performer is downgraded to an extra in the final product he is entitled to an extra session fee, but no use payments.

Outgrading: When a principal performer is completely edited out of a commercial he is entitled to no more than the initial session fee.

Upgrading: When an extra is upgraded to the status of a principal he is entitled to the session and use fees of a principal performer.

Maximum period of use: The maximum period during which a commercial may be used cannot be more than 21 months after the performer first renders employment.

Fixed cycle: Each period of 13 consecutive weeks beginning with the first day of employment, not to exceed 21 months.

Use fee: The amount of money due a performer according to the usage—how much and where a commercial runs—to be paid at the end of each 13-week cycle.

Holding fee: An amount equal to the session fee paid to the performer during cycles when the commercial is not run on air. A holding fee insures the advertiser's right to run the commercial again during the 13 week cycle, to some degree compensating the actor for holding his exclusivity, even though the commercial may not be running and accruing usage fees. Holding fees are paid at the onset of every cycle and may be credited against any use fees incurred during that same cycle. The first session fee acts as the first holding fee.

National network: Commercials broadcast on one or all of the major networks nationwide, including each of the three major markets—New York, Los Angeles and Chicago. This is not a union term, but one used widely in the industry.

Wildspot: Broadcast on noninterconnected single stations, used independently of any program, or used only on local programs.

Program commercials: Divided into Class A, B, and C. Class A connotes usage in two or more major markets, Class B in one major, and Class C in no majors.

Dealer commercials: Designated as Type A or Type B, these are commercials made and paid for by the manufacturer or distributor and then distributed to local dealers. The dealer runs the spot locally at his own expense.

Dealer: Independent company that offers a product or service for sale, a chain of local retail stores, or local outlets.

Seasonal: A commercial that is related to a particular season, such as a Christmas commercial, June bride commercial, or Valentine's Day commercial. These spots run for one cycle and do not hold the performer's exclusivity.

Cable use: A bone of contention in the industry—some people attribute the decline in residual payments to the increased use of commercials on cable. Payments in this area have been minimal at best, and disagreement on this topic was a primary cause of the SAG strike of 1988. Some major strides were made in the last contract negotiation, and for the first time cable includes a use structure. The maximum use period for cable is one year from the date of production. Exclusivity is not required on cable-use only commercials, however many commercials are provided for use on network as well as cable, with exclusivity being held in that area as usual.

Foreign use: Not including the United States, its commonwealth's territories and possessions, Canada or Mexico, foreign use is broken down as the "United Kingdom," "Europe other than the United Kingdom," "the Asian-Pacific zone," and "anywhere in the world outside of the United Kingdom, Europe, and the Asian-Pacific zone." SAG sets forth scale payments for each of these usages, but they are minimal, and agents can usually negotiate more when commercials are produced for foreign markets.

Overscale: An agent can negotiate as much as she is able for her client, such as double, or triple session fees and use payments. She may also or instead secure a guarantee.

Guarantee: An agreed upon minimum overscale amount per cycle, against which usage fees are credited. For instance, an agent may secure a $2500 guarantee, which assures the performer of at least that amount for every 13 week cycle. Any amount above and beyond the guaranteed amount also goes to the performer.

Not applied: Not a union term but still widely used, session fees that do not get applied or credited toward the usage amounts. Not-applied session fees are paid in addition to residual payments.

13th-use hold: Also not a union term, but rather a little-known negotiation tactic in which the agent gets the advertiser to agree to a 13th-use hold on residuals. The usage rates drop after the 13th usage, but putting a hold on it retains the original amount. In the final analysis this can amount to quite a bit of money.

Residuals: Usage payments.

Renegotiation: After the maximum use period has elapsed or the commercial has been dropped, it must be renegotiated before it can be reinstated, or put back on the air, provided the agent or performer has sent out the proper notification. It is at this point that the agent can often negotiate overscale payments for her client.

Reinstatement: Putting a commercial back on the air after it was previously dropped.

Working Conditions

Consecutive employment: Allows for payment on days for which the performer is not working, but fall between work days, excluding weekends and holidays.

Overtime: Time and a half for the ninth and tenth hours and double time thereafter.

Engagement: A performer is definitely engaged for a job when notified in writing, is in receipt of a contract, fitted, given a script, is told not to accept a job for a competing product, or, as is commonly the case, given a verbal call. A date does not have to be set to be definitely engaged.

Postponement and cancellations: A producer may cancel or postpone a shoot due to an act of God or *force majeure* with no penalties. Barring that eventuality, he may reschedule within 15 days of the original date, as long as the performer is notified prior to 24 hours before the original shoot. This entitles the performer to one half the session fee. If the job is canceled or postponed within 24 hours of the original date, the performer is entitled to a full session payment.

Rest period: The principal performer is entitled to a 12-hour consecutive rest period from the time he is dismissed from the set to the first call thereafter. If he chooses to waive this right, he is entitled to a $500 fee. He is also entitled to a 5-minute rest period in each hour of actual work.

Work time: Generally speaking, the period of time between reporting to work and dismissal at the end of the work day.

Makeup and hair: If other than ordinary makeup or hairdress is required by the producer, a professional must be provided to apply and maintain the makeup and hairdress.

Wardrobe: Performers who supply wardrobe are entitled to a small fee.

Meal time: Between one-half and one hour, to commence within six hours of the call to work, subsequent meals to commence within another six hours.

Casting and auditions: The performer is entitled to compensation for auditions in excess of one hour, or for three or more auditions.

Ad lib or creative session calls: When the performer is not given a script and told to "improvise," or is given a script and told to improve on it substantially, then that constitutes an ad lib or creative session, and entitles the performer to a fee.

Individual voice and photographic tests, fittings, wardrobe and makeup tests, rehearsal time: The performer is entitled to a fee for voice and photographic tests in excess of one hour. Fittings, wardrobe, and makeup tests performed on the same day of the call to work are considered work time and are to be paid for as such. Any of these tests performed prior to work are to be compensated on an hourly and subsequently one-quarter-hour basis. Rehearsal time also constitutes work time and shall be compensated as such.

Night work: Work between 8:00 PM and 6:00 AM entitles the performer to a premium of 10%.

Saturday, Sunday, and holiday work: All entitle the performer to extra session payments.

Weather permit calls: Rescheduling of outdoor shoots due to inclement weather to a subsequent date entitles the performer to a fee, the amount due depending upon when the determination is made.

Travel time: Compensation for travel time to and from a work place is computed in fractions of an hour.

Interview with John McGuire
Associate National Executive Director
Screen Actors Guild
1515 Broadway
New York, NY 10032

BEARDSLEY: How did you arrive at your present position?
MCGUIRE: My background is in legal. I am a lawyer and I was working at the Corporation Counsel's Office for the City of New York. That was 23 years ago.

It was just a chance that they [SAG] happened to have contacted me through the law school I attended, and I came in for an interview because I was looking for a change. Here I am many years later.

BEARDSLEY: And now you are the chief negotiator for SAG?

MCGUIRE: We have two chief negotiators. Ken Orsatti is the National Executive Director in the Los Angeles office. We jointly negotiate all the contracts.

BEARDSLEY: Can you tell me a little bit about the collective bargaining process?

MCGUIRE: It really starts back at the union side. Probably six to eight months before the negotiation starts, committees of actors meet and analyze what they like and don't like about the contract and what kind of changes they would like to see made. That is done all over the country and ultimately results in a set of proposals that is then coordinated, so that everybody agrees upon a particular set of proposals. That is then taken to the board. We often negotiate with AFTRA.

BEARDSLEY: AFTRA is involved in the process?

MCGUIRE: Oh yes. We negotiate together with AFTRA because they also have jurisdiction over radio and television commercials. The final proposals are taken to a meeting of the board of directors of SAG and AFTRA. Once those are approved, they are presented to management [the advertisers]. Management, I guess, has a somewhat similar structure in getting proposals together. Then there is a lengthy series of meetings. Usually they last two to three months and, when it works correctly, result in agreements. When it doesn't work correctly, it has resulted in a strike, but, fortunately, only a few times.

BEARDSLEY: Who negotiates for the other side?

MCGUIRE: There is a professional negotiator who is hired by them and negotiates for them. He represents a combination of advertisers and advertising agencies. They have a committee composed of representatives from pretty much all of the major advertising agencies and most of the major advertisers.

BEARDSLEY: How often has a negotiation ended in a strike?

MCGUIRE: We had a strike in 1978 that went over to '79, and then there was one in '88. Before that, we hadn't had a strike since the 1950s when the contract was implemented.

BEARDSLEY: There are so many nonunion actors dying to work, that when a strike happens some see it as an opportunity to get their foot in the door. What's your feeling on this?

MCGUIRE: From our standpoint, what we usually advise people is that if they are serious about becoming a professional performer, even if they are not yet a member, they should be aware of what is at stake for professional performers when there is such a thing as a strike. Therefore, they shouldn't work, they shouldn't take jobs, because ultimately they want the benefits of that contract. They will want to come into this union and work with fellow performers. We realize it is difficult because they are not in the union. We strongly advise people to avoid getting into the situation of scabbing in the event of a strike, because their fellow performers do not forget those things.

BEARDSLEY: Do you get enough support from the union members in terms of putting the proposals together and attending meetings?

MCGUIRE: That is a particular area where member involvement is very heavy. They are very anxious to get involved with making changes in the contract.

BEARDSLEY: Which areas are weak in terms of member participation?

MCGUIRE: Attending membership meetings, voting, the kind of things that I guess you find in other areas that involve participatory democracy.

BEARDSLEY: Can we talk about strengths and weaknesses of the current commercial contract? Which areas need the most work?

MCGUIRE: In any contract, there are always areas where it can be worked on, but I guess more specifically at this point, probably the ongoing area that needs the most work is cable.

BEARDSLEY: What are your thoughts on the recent changes in the cable use structure?

MCGUIRE: I think there has been substantial improvement in the amount that is paid for use on cable as well as a structure that begins to reflect the growth in cable based on the size of the various cable networks that the commercials are placed on. It still does not equal the kind of structure that we have, for example, on broadcast, which is called Class A Program Network. This requires a payment for each use. Cable is still an unlimited use—for an amount of money over a specified period.

BEARDSLEY: Are the advertisers fighting the cable use structure because they are planning to use cable more and more?

MCGUIRE: Well I am sure, from their standpoint, they are fighting it simply because they are trying to control their increased cost. Our point is that you are delivering the same number of people on cable as you deliver on broadcast and you want to be paid the same way.

BEARDSLEY: What is the attitude of the opposing side? Do they steadfastly refuse to relinquish any points, or is it more of a give and take?

MCGUIRE: It's a give and take. Nowadays, when management comes to the bargaining table, they have their whole list of things that they want to change in the contract. Many years ago, only the union had a list of things that it wanted to change. Now, management comes in with their list and a lot of it is a give and take as to what gets changed from the standpoint of the union proposals and the management proposals. But I think that the overriding goal of management is to try to compute the cost of any changes and to keep those costs within a certain overall percentage.

BEARDSLEY: What are some of the things they would like to change that we possibly are not aware of at this level?

MCGUIRE: Well, I think management has always wanted a change in how retail commercials are paid under the contract. They would like to change a lot of things that they're not likely to—for example, changing the residual structure in a more significant way. But we have to be careful in publicizing information about certain management proposals during the negotiations. It may be a proposal, for example, that one side would like to acheive but is not prepared to go to the point of a strike over it. You don't want to give such proposals more weight than they perhaps warrant.

BEARDSLEY: Other than cable, are there any other significant areas that are slated for work in future negotiations?

MCGUIRE: Let me just say this, to kind of put into perspective a contract area such as foreign use. As the world becomes more international and you have more opportunity to use the product not just within your own country but everywhere, then your foreign use payment structure becomes more and more important.

BEARDSLEY: A lot of actors cross the foreign use clause out on the back of the contract. Where do you stand on that?

MCGUIRE: Well, that gives the individual the ability to negotiate more than union minimum. The union's concern is what the minimum is, so whether the person crosses it off or not, there is, nevertheless, an established right to receive compensation; and our job is to make sure that that minimum compensation is adequate for the use. But, when they strike it off the back it simply means that the individual is reserving their right to negotiate to have something more than minimum. Or maybe, they don't want it used in a particular market and that may be the reason they would cross it off.

BEARDSLEY: Are you familiar with the personal manager situation? Actors are often unhappy with contracts they have signed, or fees they have paid, or photographers they have been "encouraged" to use through personal managers. Is there a solution on the horizon for the formal organization and regulation of personal managers the way that agents are franchised? Is there a possibility of union involvement?

MCGUIRE: In line with your comment, there are definitely problems. It would probably take an in-depth investigation to determine the full nature of the problem. I know there are very good personal managers out there who obviously do not gouge people, who do a good job, and who are valuable from the standpoint of guiding a person's career, which is what management is supposed to be all about. Then there are a lot of others who are probably violating the law because they are soliciting regular employment and should be licensed by the state. But, I guess ultimately, if there are too many of these problems, people are going to have to seek a legislative solution.

BEARDSLEY: So it rests with the state?

MCGUIRE: I mean if people are soliciting employment regularly and yet are not licensed by the state, they are obviously in violation of the state law, and that is something that should be looked into. Whether or not there should be licensing of managers I think would have to be fully debated on a state level.

BEARDSLEY: What about putting a cap on managers' fees?

MCGUIRE: There are only two ways to do that. One is through state legislation, the most common approach, the other is whether a cap is placed on agents' commissions because the union negotiated the cap as well as the state. . . but that is where the union has a right by law to regulate agents.

BEARDSLEY: Is there anything you would like to say to actors about the union?

MCGUIRE: Well, the big thing I want to emphasize is that the union tries to adapt to what is happening and not get fixed into positions because that's the way it's been in the past. It tries to understand the changes in the business, the economics of the business, and adjust the contract accordingly. The ultimate goal is always to make sure that people who otherwise cannot be adequately protected on their own will be protected because collectively they have a union.

6 □□□
□□□
□□□

The Audition and Booking Process

The audition is your chance to show the casting director, producer, director, and everyone else concerned why they should hire you. Commercial auditioning is not necessarily about great acting, although that contributes, but about portraying your type and using it to sell their product in a condensed period of time. This means creating the environment, character, and situation, and making it real for you and them in 30 seconds of the advertiser's version of the real world.

Whether a hard or soft sell, you must take hold of the copy and make it yours. Believe in the product and in the words they have written for you to say—no matter how hokey they may be. They should be as fresh, real, and believable as though no one ever said them before. As Spencer Tracy so insightfully said, the best actors never let you see them act. Do this with the right combination of enthusiasm, personality, and looks, and you will prove to be irresistible.

The Audition

If you do not already have a notebook, get one now. Keep it and a pencil on your person or by the phone at all times. Some agents routinely deliver a large amount of information, which may be quite complete, while others give only cursory or minimal information because either they don't ask enough questions themselves or they don't believe actors need to have more than the basic facts of when and where to go.

It is your right to know as much as possible before heading out for the audition. While you may not be able to get all of the following information, get as much as possible; and, when you win the callback, have them fill in the blanks.

What You Need To Know
- the name of the product

- the casting director's name
- the location of the audition including floor or room number and cross street
- the time of the audition, and in the case of a scheduling conflict, whether there is any flexibility with the time
- where and how the commercial is running
- which conflicts are being held
- the anticipated shoot date(s) and location
- the character and age range they are looking for
- what you should wear
- if there is copy, and if so, how much (it could be anywhere from one word, to some dialogue, or 60 seconds of *wall-to-wall* copy)
- which agent (if there is more than one within the agency) is sending you on the call

It helps to have other information for your reference as well, such as the name of the director, what types he habitually goes for, whether they are casting in other cities for this spot, and any other insights or inside information your agent may have. Everything should be recorded in your notebook. Whether or not you get a callback, this information could be useful in the future. If you do get a callback, it will help keep your records straight and avoid future problems.

Preparation

Hair should be well groomed but not overdone. Makeup, unless noted otherwise, should be light and worn as you would for every day. Suggest the part with your wardrobe, within the guidelines given to you. Don't rely on the decision makers having imagination or creativity; just give them what they want. If you are supposed to be acting in a black tie affair, show up in a jacket and tie; not overalls. Men should carry a razor for late day auditions. Both women and men should have their nails manicured at all times, women wearing a light or neutral shade of polish. You may be asked to hold your hands up for the camera.

Some actors suggest buying the product, if it is not expensive, beforehand. This will help you become familiar with what you are supposed to be selling, and you'll also have it on hand to use as a prop. Although casting directors usually have something to use as the product on the premises, it can't hurt to be prepared.

Remember to bring a picture and resume. Although Polaroids are now routinely taken on commercial auditions, they may also want to have your 8x10 on hand.

What To Expect, What Is Expected of You

Be sure to show up on time. If you are running late, call your agent and have her call ahead to the casting director. This can mean the difference between being seen and not seen, and will avoid bad feelings. In New York on rainy or inclement days nearly half the acting population decides to stay at home in bed. Be the one to make the effort and show up on time looking great, despite the unpleasantness outdoors. Agents are prepared for a day of aggravation when it rains. Their incoming appointments usually cancel, as well as most of those going on auditions. If the actors *don't* cancel, they show up late, assuming their lapse will be excused because of the weather. This messes up the casting sessions and sends the agents scrambling for last minute fill-ins and replacements. Although lateness on bad weather days sometimes cannot be helped, give yourself a little more travel time in inclement weather.

When you first arrive you will be asked to sign in. If you are early make sure to put in your actual appointment time as these forms are used to tabulate overtime for the unions. There may be information cards to fill in as well, asking for such incidentals as height, weight, and sizes. Do not lie, as this information may be used later for wardrobe purposes. If your age is asked, either leave it blank or give a range. This is where knowing ahead of time what age they are looking for will come in handy. If they are looking for someone in their early 30s and you are 26, put late 20s to early 30s. Someone must have thought you were the right age for the job or you wouldn't be there. Do not divulge your real age, but try not to lie either, as either situation will surely backfire on you.

If the commercial features copy there will be a script there for you to review. There should also be a *storyboard*, which depicts the commercial frame by frame and will give you a sense of what they are looking for. If you don't see either the copy or storyboard, ask.

Once you get the copy, go over it a few times to get a feel for it and to work on any pronunciation. You can try to find a quiet spot, either in the hallway or bathroom to work on your lines, but don't go too far. Let the receptionist know if you leave the room. Most casting directors prefer that you do not memorize the copy, unless it is only one or two lines. They would rather you concentrate on your acting. Experienced actors concur on this. They believe that if you try to memorize it, the bulk of your energy will be spent trying to recall the lines instead of focusing on the more important task of acting. In any case, memorization simply isn't necessary. Cue cards are provided for this purpose. What they are looking for is an overall feeling.

A Polaroid may be taken either in or out of the audition room. Ask if they would like your picture and resume as well.

When you are called in, introduce yourself to the casting director or assistant running the session, but leave it at that. Idle chit chat wastes time and often comes across as inane or inappropriate. There may be other people in the room who may or may not be introduced to you. Most likely they are agency people. It is rare for the director to be present at the audition. If the people present seem receptive, shake their hands. Make sure to meet everyone's eyes. Proceed to the audition area, which will be marked with white tape known as the *mark*. The mark indicates where you should stand, in order to be in sight of the camera. You should ask the casting director how much of you will be in frame.

At this point the casting director will offer direction on what they want to see from you. Do not be afraid to ask for clarification if you do not understand the instructions. Or, they may wish to withhold direction and see what your instincts are. Do not be afraid to follow your impulses and make decisions based on your interpretation of the copy. It is better to make incorrect choices than no choices at all. Direction may be offered after you've read the copy once through. Follow the direction and make the necessary changes. Casting directors hate it when actors look like they're listening and absorbing the instructions, then go ahead and read the copy exactly as before. This is another reason you shouldn't memorize beforehand. If you memorize, some directors feel that you can't help but make rehearsed and perhaps inappropriate choices, which can also hinder you from making directional changes. Whatever the case, listen and incorporate the changes into your work.

Before you deliver the copy, you will be asked to *slate*. All that is required is to say your name. Do not read too much into this. You can liven it up with "Hello, my name is...," or "Hi, my name is...," but don't get cute and don't overdo it. Remember that all the auditions will be viewed back to back on a video reel. Several slates in a row of "Hi, my name is..." with an artificial cheeriness can get pretty tiresome. Keep it simple and natural.

Once you are excused, say your thank yous, and mention to the casting director any important recent career developments, such as going up for the lead on a Broadway play or having a meeting with the director for a feature film. This is ammunition she may be able to use to sell you to her client.

You will be asked to enter your out time before leaving the premises. If the audition is for a union job and you have been there more than an hour, you are entitled to monetary compensation. Infractions, especially if they happen repeatedly with the same casting director, should be reported. Any claim made to the union will remain anonymous unless the actor gives permission for his name to be used. The casting director does not have the option of announcing that she "is not paying overtime, so anyone not liking it can leave," as an occasional casting director is known to do.

Audition Do's and Don'ts

Do not crash auditions. When some actors run across an audition they'd really like to go up on, they just show up and sign in as though they have an appointment. As sessions are sometimes run by people other than the person who set it up, this will occasionally go unnoticed until it is too late and the audition tape is already off to the client. Some casting directors are only too happy to hold a grudge and will never again see an actor who crashed one of their sessions. If you learn about an audition you think you are right for, call your agent. There may be a reason you weren't given an appointment for the job. You may have a conflict, it may not be a good job, the casting director may have turned you down, or they may be seeing people by request only. You even may have an appointment for another day but haven't received the call yet. Otherwise, the agent may be able to get you in on the session. If there is no agent with which you have a healthy relationship, tell the person running the session that you don't have an appointment, but ask if they would consider seeing you for the job anyway. This will at least give them the chance to wipe you off the audition tape if you weren't right. If you know about the audition a day or two beforehand, drop off a picture at the casting director's office and ask.

If there will be copy for you to read, show up early. Do not waste time socializing with other actors in the waiting room. When you walk through the doors of the audition room, you should be completely prepared and in charac-ter. Do not offer comments on what you think of the copy. For all you know the casting director's brother may have written it.

If you don't look like everyone else in the waiting room don't worry. You may be there to offer a "range" or in case they decide to go the "other way." If you do look like everyone else in the room, don't worry. Everyone has something different to offer, and they may be looking for a special nuance or appeal that only you can give.

Do not critique your performance with words, expressions, or gestures, either voluntary or involuntary. Train yourself to deliver the copy and to smile pleasantly, no matter what you think of your performance. If you think you can do it better or differently, ask to try again.

Do not be nervous. Be in control of the situation. Do not mumble, but speak clearly and distinctly. If they don't know what you are selling, it doesn't matter how well you act. Don't ever forget that the product is the primary focus of attention.

Keep your actions minimal and within the range of the camera. If you start an action, finish it. Be deliberate.

Do not deliver your lines to the wall or to people in the room, but to the camera. Be intimate with the camera, imagine that this is the person you are speaking to. Know whom you are addressing, the situation at hand and where you are supposed to be. This will come through in your performance. Keep your tone conversational. Be consistent with the rhythm. Do not plan how you are going to deliver your lines beforehand—keep it fresh and natural. Be energetic and enthusiastic without going overboard. Do not overact. Be real.

Do not hesitate to follow your impulses or take a risk. Even if your choices turn out to be wrong for that particular spot, the casting director will remember who you are. Hopefully she will be favorably impressed and will call you in again. Do not try to be like everyone else. Try to let the real you shine through.

Use humor. Do not be a cold fish. As discussed in Chapter 1, everyone likes to smile. This does not mean to stand there and tell jokes or make light of the copy. Incorporate humor wherever and whenever you can, without being stilted or artificial.

Once again, never let them see you act. Keep it light, natural and fresh.

The Callback

Receiving a callback means that you are being considered for the job. The principals involved may want to see more of what you can do, convince other parties that you are right for the job, or see how well you mesh with the other actors under consideration. At this time they also may ask you to hold your time for them in the form of a *first refusal* or *general hold*.

The First Refusal

Receiving a first refusal means that the advertiser wishes to have first rights to your services on a particular date or dates and wants to keep you from working on a competitive product until they've decided whether to book you or release you. Not recognized by the union, first refusals have nonetheless become an industry standard and are a bone of contention for performers and their agents. The beginning performer may not see a problem with allowing her time to be reserved, but as her popularity grows and more products and dates are put on first refusal, the situation can become quite entangled. The performer is not reimbursed for holding this time, and many others may in fact be holding dates for the same role. What happens more often than not is that the actor ends up with no job at all. Since casting directors are obliged to exact first refusals for their clients, those actors or talent agencies refusing to comply would simply be left out. First refusals, according to the union, are not a contractual obligation, and Elinor London, who heads the Commercial Contracts Department in the New York office of SAG, says that the union has put advertising agencies and their casting agents on notice for abusing this practice. Lately, the complaint level has dropped off.

The Information You Will Need

At this point, you should know all the information listed earlier in the audition section. Instead of just knowing the shoot dates, however, you need to know whether you are actually holding time for them and giving them either a first or second refusal. You may also need to clarify the *conflicts* or *area of exclusivity*.

If you have even a potential conflict with the product or date, the time to bring it up with your agent is *before* the callback, not after. If you suspect that you have such a conflict, double check with your agent. Do not assume that they have the situation under control. Occasionally an agent blunders and books a performer on a spot for which they already have a conflict. If this should happen, remember that *you*, not the agent, will be held legally responsible.

If you are already holding for another job, you will be giving a *second refusal*. If you are holding for five other jobs on this same date, you will give each of them a second refusal. All a second refusal means is that this advertiser is not first in line. There is no need to upset anyone by telling them they are sixth. This is where a good agent is especially important. She will finesse the situation to see that you don't antagonize anyone or miss out on any jobs, if possible.

At this time you should also double check your union status, making sure you are paid up; and, if it is an out-of-the-country shoot, check your passport. If the shoot involves driving a car or operating some other vehicle, make sure you have an appropriate and current license.

What To Wear

Most agents instruct their clients to wear the same outfit they wore on the audition to the callback, so as to not disturb a proven formula. They may also be secretly afraid that the decision makers are too incompetent to recognize or feel the same about the actor in a different outfit. No one ever accused agents of not being paranoid. Many casting directors, who innocently chalk up the performer's tendency to wear the same outfit to superstition, feel that it is okay to wear different clothes. If this is what you elect to do, it is probably best not to stray too far from what you originally wore in case the agents are right.

What To Expect

On first or subsequent callbacks there will most certainly be several people there to watch your performance. They may include the agency producer, members of the creative team, the client, and the director. If the director is there he will be the one to direct (or redirect) your performance; otherwise the casting director will be the one to offer you additional guidance.

You may be asked to stand with several different actors, or to try different ways of working with the copy. Whatever the case, if you've made it this far you can be assured that at least someone there likes you. Now it may boil down to a matter of looks, physical compatibility with the other performers, or something else you have no control over. If you do not win the job, try not to blame yourself. It probably has nothing whatsoever to do with you. If you do get the job, they may let you know right there on the spot, or you may receive a call later from your agent. If they tell you they want you for the job, wait for the confirmation from your agent to make sure you are really booked. If you leave the premises and start receiving calls from wardrobe people or the stylist, chances are you have the job, but refrain from giving out any information until you have a firm booking. SAG does not allow size information to be given out to the client until the booking is firm. The actor may unwittingly give out his sizes only to have the client reply with, "Really, *that* big?" and lose the job to the agency's second choice.

The Booking

Even if you received some of this information earlier, double check to make sure nothing has changed:

booking date
weather permit day(s), if it is an outdoor shoot
the location of the shoot
directions or travel arrangements
full name of the product or service being advertised
exclusivity/which conflicts are being held
how the spot is running
advertising agency name
director
agency producer
AD (assistant director)
call time (they usually don't know this until the day before the shoot)
any special contract provisions

Preparation

If it is an out-of-town shoot, you will need travel information including mode of transportation, per diem, and where you will be staying. If it is a union shoot, you are entitled to $50 a day and first-class accommodations and travel. Some actors like to cash in their first-class tickets for coach and pocket the difference. If you do this, try not to let anyone know. While it is not expressly forbidden, it is looked down upon by producers who are required to pay for the first class arrangements. If you want to travel on a particular airline for frequent flier mileage or other reasons, bring it up with your agent when you are being booked. The employer is under no obligation to oblige you, although most times they do try. Most importantly, for out-of-town shoots, *get after-hours phone numbers of the contact people in the area before leaving town.* You should also bring your agent's home phone number for emergencies.

Now it's time to memorize your lines. Your agent will make arrangements for you to get the script ahead of time.

The wardrobe person or stylist will be calling you to get your sizes and work out your wardrobe. He will probably want to know what you own and ask you to bring a few things. Make sure these items are clean and pressed, and try to bring a selection. Bringing any of your own wardrobe entitles you to a fee of $15. If you are asked to bring evening clothes, the fee is $25. If it is a high budget shoot or a job that requires special clothing, they will go shopping for you. A fitting will be required. For this you will be reimbursed at an hourly rate based on your session fee. If you are making more than double scale for the shoot, they are not required to pay you for the fitting.

You will be told to wear your hair and makeup as usual, or a stylist will be provided for you on the set. In either case, it is prudent for women to bring their makeup essentials with them to the set. You may be asked to get a manicure before the shoot, in which case you will be reimbursed for the expense, but not your time.

According to Federal immigration laws, *all* performers are required to produce proof of citizenship or resident work status on the set. Called "I-9" information, it consists of either one item from list A or one from both list B and list C:

A   U.S. passport
    certificate of U.S. citizenship
    certificate of U.S. naturalization
    current foreign passport with attached employment authorization
    alien registration card with photo
B   state-issued driver's license or ID with photo
    U.S. military ID card
    school ID with photo
    voter registration card
C   original Social Security card
    U.S. birth certificate
    current INS employment registration form

The Shoot

When you arrive on the set, most likely it will be a hub of activity. You'll find the production staff, agency people, the director and her assistant(s), the camera crew, and various other busy people. Find the AD (assistant director), whose name you should already know, and introduce yourself. He will take you around to meet the director and make sure you go where you are supposed to go.

You will be given a contract on the set. If you are called in front of the camera before thoroughly going over it, put it aside until you have a free moment. Do not be hurried into signing before reading it carefully. If there is any question or dispute, do not sign it or bring up the problem there on the set, but ask to use a phone and call your agent. It may be a simple oversight or misunderstanding, or it could be that somebody is trying to get away with something. Let your agent work it out.

Be prepared to spend a full day on the set. Shoots rarely go quickly, and much of your time will be spent waiting. Try to use idle time to your advantage—either working on lines or getting to know the people you are working with and jotting down their names.

There will be food provided at various times during the day. As these are sometimes sumptuous displays of delicious but unhealthy or fattening foods, try not to gorge yourself. Remember that you will be on the set for at least a full day, maybe more. Try to avoid foods that give you highs or lows or affect you adversely.

If the spot requires you to eat on camera, there will be a recepticle located on the set, unglamorously known as the *spit bucket*. After a few bites of the sponsor's microwave hamburger or artificially sweetened ice cream substitute, you and this bucket should be very well acquainted. The bucket is put there for a reason. Do not be afraid to use it. You may be required to bite and smile for hours on end, but no one expects you to swallow.

Before leaving the set for the day, make sure to thank everyone involved. Make arrangements with the AD to receive a copy of the commercial. Follow up with thank you notes to the key people. You are likely to run into some of them again. You want them to remember you favorably.

Interview with John Massey
Commercial Director
Rick Levine Productions
59 E. 82nd St., New York, NY 10028
9026 Melrose Ave, Los Angeles, CA 90069

BEARDSLEY: How do you like directing commercials?

MASSEY: Oh, it is terrific, challenging and a lot of fun. It's very difficult, and I find it very satisfying to achieve something in 30 or 60 seconds. The challenge is to communicate ideas and concepts in that short period of time and to make it good and worthwhile to look at, going beyond just trying to give it a little bit of life and imbuing a little bit of . . . something special.

BEARDSLEY: Do you have a preference as to the type of actors you use?

MASSEY: I always go with the best actor. The first audition, is videotaped in the casting director's studio with maybe five minutes to read the copy followed by an opportunity to talk about themselves a little. I can find the essence of the actor in just that little interview. I try in the callback to find out exactly what their range is by giving them a myriad of things to go through. Frequently on the set, the client or agency says, "Can they read it this way or that way?" They have to be flexible in their performance and able to color things in many different ways.

BEARDSLEY: Do you like using the same talent on different jobs that you direct?

MASSEY: I have directed some ongoing commercials where the characters haven't changed. In these I have worked with that same talent. But most of all I like starting fresh to get a fresh eye. I know that there are some directors who have a kind of repertory going, and I think that is great.

BEARDSLEY: Who do you use for casting?

MASSEY: In Los Angeles, I like Sheila Manning; and in New York, I have used Joy Weber and Beth Melsky. Joan Lynn is really terrific, too. She has been doing a lot of my casting recently. I like the actors these casting directors bring in. That is their talent. Joan, for example, is very up on theatre. She really knows what is happening and brings in the up-and-coming good actors—people you wouldn't think of putting in TV commercials.

BEARDSLEY: Even when they don't get the job, will you work with some of these actors again?

MASSEY: I try to keep a log. I keep a mental note or even catalog people who I think have been interesting to see. If I have to do something very quickly, then I can say, "Let's get these people in to audition."

BEARDSLEY: Do you find any differences with the talent or types of spots on either coast?

MASSEY: I have luck on both coasts. Actors are mobile. They travel back and forth between the two coasts depending on where auditions are happening. If we haven't come up with someone who is interesting or people we like in New

York, then we can go to California and hit terrific people, and vice versa. I think it depends on what is going on during the pilot season out in Los Angeles—a large contingent of New York people go out there.

I hear a lot about a "California look" or a "New York look." But you can be in California and have a client who wants the New York look or be in New York trying to capture the California look. The country is getting smaller.

BEARDSLEY: Some people feel that L.A. relies too heavily on the look and not on the acting. Do you think that is valid?

MASSEY: I can't speak for other directors, but I like to approach casting looking for the best actors for all parts. I am sure there is a nugget of truth in there, and advertising does depend on a look. I think that is true overall, not just on one coast or another. Actors should not be discouraged by the subjective nature of the business.

BEARDSLEY: Among other things, you have done the Maxwell House series—a dramatic, award-winning series. Do you have a preference for doing comedy or drama?

MASSEY: I try to do as many different things as I can. I do as much people stuff as I can, whether it be comedy, drama, or whatever. I love the opportunity to try something new.

BEARDSLEY: What is the difference between casting a comedic and a dramatic spot?

MASSEY: There are actors who have a comedic bent, just through their looks or presentation. Although I know comedy takes great acting, certain actors possess a twinkle and are just naturally funny or they perceive things in a funny way. I think the same goes for more serious actors who have very serious looks.

BEARDSLEY: So it comes down to a look?

MASSEY: The look backed up by the acting.

BEARDSLEY: What do you think of the hand-held camera, *cinema verite*, and other trends?

MASSEY: I think they are terrific. These trends have done a lot to move communication forward in commercials. We are looking for ways to get visual ideas across very quickly. To a certain extent, it is a combination of not only the production but also the post production. A lot of the editorial work has lent itself to that *cinema verite* style and has come together to give you the television commercial. Sometimes that involves catching the actor off guard and getting a look or an eye movement or something and using that piece editorially. So the director or the editor at that time really becomes a creator of a visual montage.

BEARDSLEY: Do you see in advance what is coming up in the way of trends?

MASSEY: I see special techniques that have actually led to some of the work in features. I think that some of the work that Oliver Stone did in *JFK* could be attributed to visual styles that have been used in commercials in the past few years. Some of that comes from post production, editors' ways of translating and editing scenes. Much of that has been borrowed from MTV. There are so many different choices now. Maybe that is the new trend, not trying to be just one thing. People will realize that the *cinema verite* style is good for some things, a more romanticized execution is better for others, and it is the director's job to be versatile in all the different languages he can.

BEARDSLEY: Do you have any advice for actors?

MASSEY: Try to not go into the commercial audition with any kind of precon-ceived notions of what a *commercial* is. I get a lot of actors who are really concerned with how they hold the product and, personally, I don't care about that. I think that is the least of their problems. That distracts from the believability of their performance and will hurt them. They come off more as a pitch person than a natural, believable person within this kind of unreal world. Don't think about how you are supposed to be because you are up for a commercial. Try to strip it all away and be as natural and as much yourself as possible. Try not to get discouraged. It must be terribly discouraging with the decision process so convoluted, how a person is actually chosen. Just stick to your talent and what you think is a good interpretation of the dialogue.

BEARDSLEY: What *about* the decision process? Do you, as the director, have the final say, or does the client have the last word? How much input does the producer have?

MASSEY: The way the process usually works is that the production company, along with the agency, is given a set of casting specifications that have been talked over with the client before the director even sees it. It is these character specifications that we follow and give to the casting director. We are (hope-fully) in sync with what the client is looking for and the agency created. When we see that person, everyone should feel comfortable in the direction that we are going. If that changes, and we feel it's necessary to go in a different direction, we have to be prepared from a conceptual point of view to explain it to the client. The decision process in terms of talent is pretty much between the director and the agency. The director can generally get the talent through that he would like to put into the commercial, unless they think that you are way off base, in terms of what and who they want to be representing their product. You have to be able to turn to that client and say, "Well this is really why we are looking at this particular person because they give you this . . . " The clients are getting pretty tuned in. But, ultimately, they have the final word.

BEARDSLEY: Are the character specifications usually enough to go on?

MASSEY: It depends on how exact these specs are. They can be "40-year-old woman," or they can get narrower and narrower. The specs are generally quite broad, and it is up to us to find the talent and sell the talent, showing to the client that they are good actors and capable of delivering what we are after. It all comes down to selling our ideas. The whole business revolves around that aspect, selling at every level.

BEARDSLEY: How often is your first choice the one booked for the commercial?

MASSEY: 75% of the time our first choice is the one booked, but you should be happy with your second choice, too. This has always been my philosophy, whether on the directing or agency side. If you don't have someone for second choice that you are happy with, then you must go to the client and say, "Look, there *is* nobody else. You have to trust us on this." There is no sense in recommending someone who ultimately you are not going to be happy with.

BEARDSLEY: Do you and the producer always see eye to eye? How does the deliberation process work?

MASSEY: It is a difficult process, at least I find that it is difficult. You weigh the pluses and minuses and give it a lot of thought. A lot of searching and looking goes into it. When you sit down and initially go through tapes, you are screening down to a list of possibly ten out of 300. By that time you should generally be pretty much in agreement. But if in the casting process they don't like the people who keep coming up, it's time to say, "Let's talk about this concept; we're seeing different people in this." It becomes more of a conceptual discussion and the differences then usually get worked out.

BEARDSLEY: Do you attribute any importance to the way an actor does his slate?

MASSEY: An actor shouldn't be goofy. Again, try not to project yourself as something you are not. Don't try to be too funny or too cute. That's my kind of turn off. I look for some honesty, and fooling around on the slate never shows me anything. I take it as a negative.

BEARDSLEY: What's the difference as you see it between acting for the camera and acting for the stage?

MASSEY: I have a great love for actors; I think they are terrific. They put themselves out there in the most amazing way. The camera tells the truth. Any kind of false move or dishonest gesture becomes magnified. You can't hide behind technique as much as you can on stage. The camera gets right to the heart. I could never do it; it is so intimidating. There is something about stage acting that is a lot different from acting in front of a camera. Some actors have an innate skill for it. In working with celebrities, you see them in real life, but when they get in front of the camera you say, "Oh, my goodness." Looking through the lens at them you realize that they are extraordinary. You can really see an inner something.

Theatre is all about projecting outward, talking outward, talking out. Someone said, "A really good actor in film draws the camera to him." I think that is the basic difference between film and the theatre.

BEARDSLEY: What is your feeling about the first-refusal/callback situation that some believe has gotten out of hand in New York?

MASSEY: In the initial casting process the interpretation may come from the casting director with a little input from the director over the telephone. Then, if on the audition reel you say, "Gee, this person looks really great, but I don't know if his acting talent is up to it," you are going to call that person back in and work with him a little. If I am casting a commercial, I wouldn't put 30 people on first refusal. Maybe it is happening on my jobs and I don't know it, but when I present a talent to the client—a first choice and a back up—both must be on hold. What would happen if you are locked into your shoot days? The talent should say, "Next Thursday, I am available to shoot your job." It should be down to one or two at that point, not ten people on hold for one part.

Even though it may be discouraging, actors should look at callbacks as more of an opportunity to seen, as a possibility to be cast the next time, if not on that particular job.

BEARDSLEY: What happens if the talent needs an answer before you are ready to give it?

MASSEY: I know that on one occasion I have booked a talent because the client would not have been able to make the decision then, and we had to book them or lose them. If the job is canceled, you eat the cost. It is a major decision you are up against. I have never put ten or 12 people on first hold. But after the call-back I would think that either the casting agent or the agent should be able to find out if they are seriously being considered.

Actors do have a point. I think there is a better way of doing things, and probably it means not calling back so many actors. I tend maybe to call back more than necessary, because I think, if there is a spark there—something there that I see—why not give the person a chance to get something interesting on tape? That is how I approach it. Now, that might not look fair to the actor. I can certainly understand that. But, that is just some of my craziness. I will pull someone out if they do something I think is interesting. I'll call them back. I will not just brush past people, which I think is an important part of the job.

There are a lot of directors who do call back more than they need. Maybe some of the actors [are put off by it] because they are not as connected as they would be with a film, feature, or theatre director. There may be a stronger connection there, because the actor comes back, reads the scene, is in there for half a day. Whereas in the commercial, she comes in, gets ten minutes, and is asked to leave. It must seem very distant and unfriendly. Actors might misinterpret that as coldness. You have to leave all that outside. Just recognize the process for what it is.

A former producer for Ogilvy and Mather, John Massey has won awards and recognition for his directorial work. Some of his commercials include Maxwell House, Phillips Milk of Magnesia, American Express, Stanley Tools, Citibank, Milk Advisory Board, Blue Cross, Jello, and J.C. Penney.

7 Children in Commercials

Most parents at one time or another consider the idea of getting their child into show business. After so many choruses of, "He's so cute, he could be in the movies," many do take steps to put these thoughts into action. Some have seen their efforts rewarded while others have been mightily disappointed. Unfortunately many have lost quite a bit of money that they need not have spent. Success in the child acting area should not initially require large outlays of cash. What it does require is a sincere desire on the part of your child to be in this business, a clear parental understanding about what is involved, and dedication, levelheadedness, and follow-through on the part of both the parents and child.

What Is Involved

Both child and parent must make a tremendous time commitment. You cannot succeed in this business, nor any other for that matter, with a half-hearted effort. No agent will begin working with a child who only wants to do certain jobs or who only wants to go out once in a while, unless that child has already made a name for himself.

Auditions for school-age children are held in the afternoon so as not to disrupt too much of the child's education. Auditions for babies and toddlers may be held anytime. Generally, about 24 hours advance notice is given. There may be hundreds of children at a single call or only a handful. Infants and small children are booked in multiples, usually five or six per role—so that if one doesn't work out because of temperament, illness, or other reasons, another takes her place. Although all the children booked command session fees, only the one used in the final commercial gets residuals. Older children are booked with a backup. The backup is paid a session fee to spend the day on the set, to be used only if the first child needs to be replaced.

A child gets paid the same as an adult, so on a very successful commercial he may make as much as $30,000 or more. However, getting to that point can take months or years. Most commercials net the actor a few thousand dollars each.

Stories abound about kids booking a job on their first time out. While this does happen, and more so with kids than adults, it still is not the norm.

What It Takes

The children who work are the ones truly who are ingenuous, with wide-eyed, innocent appeal. They are precocious without being jaded, intelligent without being hardheaded, talkative and personable without being calculated. In short, the children in demand are truly children, not little adults. Impossible, some say, for as soon as a child starts working he loses his very "childness" by sheer virtue of the fact that he *is* working, and not out playing baseball or doing his homework like his compatriots are. A child actor may in a single work day reap a salary bigger than the combined annual income of his parents. He is dealing with adults day in and day out, some of them CEOs of companies or world-renowned directors—a situation intimidating to most adults.

It is entirely up to the parents to help the child walk the tightrope between childlike innocence and the worldliness of adulthood, not simply for the sake of retaining her employability, but to keep her from growing up before her time and missing out on childhood. To do this requires maintaining a skillful balance of all the things the child finds enjoyable, whether it be show business, dance class, sports practice, participating in chess matches, or simply playing with friends. While initially it may be prudent to make the child completely available in order to convince an agent of your commitment, the child's best interests dictate that his other activities should soon be restored to his schedule as soon as possible. This involves setting priorities and missing out on a job or two. Remember, this is supposed to be fun for your child. When it starts to become *work*, it's time to stop.

It is up to the parent to help her child keep things in perspective and not allow him to think himself different, special, or better than other kids his age. It is the parent's role to keep her child from feeling the pressure of "having to get the job," or the depression of rejection. The child who feels these things is usually echoing the feelings of the parent. Most kids left to their own devices feel little regret over losing a job or superiority about making a tremendous amount of money.

The parent has to be ready and willing to go to auditions at the drop of the hat, possibly moving to another city for the sake of the child's career, willing to invest money in photos and lessons when the time is right, and ready to pull the child entirely from the business if it appears to be affecting him adversely, regardless of the time and money invested. Parents have to look within, and make sure that a show business career is being pursued because the child wants to, and not to satisfy their own needs. It is truly the selfless, knowing, and loving parent who is able to help a child gain success as an individual, regardless of whether that success is in or out of this business.

What They Are Looking For

As mentioned before, innocence sells. In addition to a freshness and childlike view of the world, a child needs to bubble over with personality. Agents and managers say they like to see a child who can talk endlessly, showing enthusiasm and glee over even the smallest things. Like adults, the child destined for success must possess a certain something special—an indefinable aura or charm that causes people to stop and take notice.

Natural acting ability is a must. All the acting classes in the world cannot instill in a child the ability to transmit meaning and personality in a glance or spoken word.

Virtually all kids are cute, especially very young ones, so unless they are extremely unattractive, most children who possess freshness, talent, and personality can work. Of course, all-American, blue-eyed blonde types are still very salable in commercials; but, as with adults, there is a growing market for character and ethnic types.

The child must be very independent, and able to function in the world of adults. Even babies must be independent and not afraid of separation from their mothers. They must be able to turn on the charm for strangers. Babies who work are still the traditional Gerber type: big eyed, chubby-cheeked gurglers.

Getting Representation

Children who are just starting out do not need professional photos. Take several rolls of pictures and make duplicates of the best ones. Select photos that are sharply focused and clearly show your child's face. Call your local SAG office and get a list of union-franchised agents in your area. Get a list of managers from the East or West Coast office of the National Conference of Personal Managers. This fledgling organization does not represent all the worthwhile managers in the nation, but you can at least be assured that the managers who do belong adhere to the conference's guidelines and thus are relatively legitimate. (Please read the sections on agents and managers in Chapter 4.) Before mailing photos, make sure that each one is labeled on the back with the child's name, parent's name, address, phone number, date of birth (not age), physical characteristics such as hair and eye color, weight, and height, and any special abilities. For babies this means anything that's not the norm, such as standing, walking, or talking at an early age. For older children this means skills such as skateboarding, dancing, playing an instrument, or speaking another language. Type the information on pressure-sensitive labels and adhere them to the photos. Do not write on the photos with a pen.

Most agents will keep pictures of babies and very young children on file and only will call you in if there's an audition. Older children will need to interview. This is a very simple procedure that will allow the agent to get to know you and your child and evaluate the child's motivation and ability to work in this business.

Representation by a manager is necessary if you cannot get an agent interested in your child or if you live in Los Angeles. Agents are essential in getting auditions. Managers are helpful in getting in to see agents who otherwise would not see your child and in offering more individualized attention. On the West Coast, agents deal through managers in the child area. If you can get a franchised agent on the East Coast to work for your child, it may not be necessary to engage a manager, unless you feel overwhelmed by all the attention your child is getting and need more personal guidance than what the agent is giving you.

Money Matters

Too many parents spend much more than they need to get their children started in show business. Because of a lack of expertise in this field, parents are particularly susceptible to unscrupulous types eager to take advantage.

How To Avoid Scams and Rip-Offs

Never under any circumstances pay money to anyone up front. Agents who are licensed by the state and franchised by the unions as they are required to be are not allowed to charge you anything. They can take 10%, and no more, of your child's gross earnings once he begins to work. Managers, on the other hand, do not have to be licensed or franchised, so the territory is much more tricky. They charge anywhere from 10% to 25% of your child's earnings above and beyond the agent's fee. Legitimate agents and managers *never* ask for money up front.

There are many organizations, some of whose practices have been examined on news programs, that present themselves as agents or managers and charge enormous fees to include your child's picture in impressive looking books that are sent to agents and managers. *These books are a waste of money*. At best, the agents who receive them scoff at the amateurishness of the photos before tossing them into the garbage. These organizations purport to have as clients well known and enticing companies, which they name in their literature. They know full well that parents who try to contact these organizations to verify the claims will probably not be able to get through to a person who can help them. Once you pay your money to these organizations, you will never hear from them again. *This is not the way to find work or representation.*

Some managers, legitimate or not, will have their clients sign what amount to unconscionable contracts. Particularly common among managers located outside the primary markets of New York, Los Angeles, and Chicago is a clause in their contract that entitles them to a percentage of their client's income long after they've ostensibly parted company. Many times a child goes on to become successful, only to hear from a manager she hasn't seen in years, looking for his share. Just as appalling are the clauses favored by the people who run pageants and conventions and by the managers who suggest their clients enter them. Such clauses entitle them to money the child (or adult) brings in as the result of meeting agents in the city or cities the pageant is traveling to. This is after the parents have already paid the pageant representatives handsomely for the privilege of having their child compete in the production, as well as having spent a tremendous amount of money for costumes and travel. Even if the parent sets up a meeting of her own accord in the city the pageant is showing in, not as the result of having been seen in their production, the child still must pay the pageant representatives and/or the manager a percentage. *Never sign anything other than a standard union contract without having a lawyer go over it first.*

Some managers, who in other respects may be quite legitimate, require potential clients to take a particular class or use a certain photographer before accepting them as clients. While this is not necessarily a scam, it may be a rip-off. Ask for a selection of classes and photographers if you desire to pursue that avenue, but look elsewhere if they insist you use theirs. This is a sure tip-off that they have a vested interest and do not have your or your child's best interests at heart.

Parents are easy prey, all too eager to believe that their kid has something special. People knowing this go so far as to obtain mailing lists from obstetricians and hospitals and send flattering letters to people who have just had babies. Parents will fork over a tremendous amount of money they can ill afford to spend, if it means getting their precious child into the glamorous world of show business. *There are no guarantees or quick and easy routes to success in this business.* Anyone who tells you so is either a fool or a liar.

When To Spend Money

Scrupulous agents and managers who know the business will tell you that they would like to work with your child and see what happens. If it happens that the child likes it and is good at it, *then* you can sign a contract and invest in pictures or classes.

Pictures

Most agents and managers agree that professional pictures are not necessary until the child begins to get callbacks and to book jobs, and even then they aren't essential until the child is several years old. Until that time, update your child's snapshots every six months, and be diligent about sending them out, even to people you haven't heard back from. Professional shots generally cost between $150 and $250, depending on how many rolls the photographer shoots. They should be natural and real, presenting your child as he really is, not gussied up and looking like a little adult. (Reread the section on photographs in Chapter 2. These guidelines apply here as well.)

Classes

Some classes may be necessary, but be careful of overloading your child. There is considerable opinion in the industry that acting classes destroy a child's natural approach to acting, and are unnecessary, if not downright harmful. Speech classes may be called for if your child has an accent, and dance classes may be in order if your child has a natural bent and desire to pursue that avenue. Be wary of singing lessons, however, which have a tendency to override the child's natural voice and give her a stylized and practiced approach. Your agent or manager should be able to guide you in this area. Seminars given to educate the parent in the ways of the business, held by agents or organizations that refer promising children on, may be worth the money if they are in fact legitimate. Find out who is running it, what their

Figure 8.1 DANIELLE KUHNS
Photographer—Linda Kuhns
Full of life and personality

Figure 8.2 MELISSA HART
Compelling eyes

Figure 8.3 GRAHAM KOLBEINS
Photographer—Paul Sirochman
Personality and charm

Figure 8.4a ALEXANDRA BEARDSLEY
Acceptable snapshot to reproduce and send out

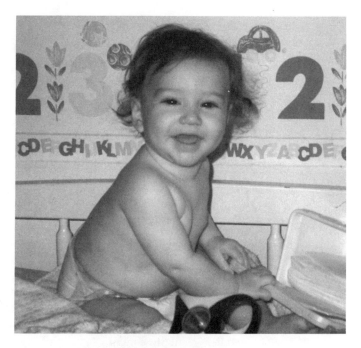

Figure 8.4b NICHOLAS BEARDSLEY
Acceptable snapshot to reproduce and send out

affiliation is, and the names of children they have discovered or referred before. You can also check with the Better Business Bureau and State Attorney's Office to determine if complaints have been filed or investigations are in progress.

Books

Many managers and print agencies have *books*, or *head sheets*, that they ask their clients to be part of. These are sent around to agents and casting directors. Once you have determined that the agency or manager is legitimate, it can't hurt to be on their book or head sheet. It should cost in the neighborhood of $50 annually, which defrays printing and mailing costs. Exposure of this sort could help get your child known and working.

Pageants and Conventions

As mentioned before, pageants and conventions generally cost a tremendous amount of money, not only in entrance fees which can amount to several hundreds of dollars, but also in travel costs for the contestant and parent and for the increasingly elaborate costumes required. In addition, regular "pageant kids" tend to have a very adult and practiced air about them and as a result they don't often book acting jobs. However, signing up with a local pageant that is coming to one of the big cities is not a bad way of testing the waters. Some pageants do enlist a number of reputable agents and managers to judge their entrants, and many young people have signed with agencies through this venue. If you want to give it a try, check to find out which agents the pageant guarantees will be there, then check with your local SAG office (see Appendix) to make sure they are franchised. Then call those agents and find out if the pageant's claims are true. Only enter one division of the pageant, preferably talent, as this is the category that most of the agents prefer to attend. This will help keep the cost down. Your child does not have to win to catch an agent's eye. Don't sign anything you could regret in the future. Before the pageant, contact the agents in the city you will be visiting and try to set up meetings. Participating in one or two well selected events of this sort will not ruin your child for show business, but making a habit of them might certainly do so.

Whenever preparing to spend money, whether it be for pictures, a manager's book, a pageant, or an organization that sends out your child's picture for you, check with a SAG franchised agency to see if they can tell you if the expenditure is worthwhile. You do not have to be a client to get this type of information. Most agents are only too happy to help keep parents from getting ripped off, and would dearly love to put an end to the unscrupulous practices that reflect on the business as a whole. If you can't get the information you need on the first call, try another agency.

Preparing for Interviews and Auditions

Most agents and managers will want to talk with the child alone, away from the parent. This allows them to see how the child functions without the parent at his side, and to determine if this is something he really wants to do. If this goes well, the parent will be called in to discuss the matter further. The child may be called upon to read a line or two of copy, or to talk about something that is special to him. They are looking for natural enthusiasm and energy.

On auditions, the child will be separated from the parent as well. If your baby is at a stage where she screams when she can't see mommy, don't waste your time and theirs. Babies and children are very unpredictable, and anyone who works with them is very well aware of the fact. It's to be expected if your child occasionally doesn't perform well or behave himself. If she consistently acts up, however, it may not be the right time to pursue this career.

When giving out the audition information, the agent will let you know how the child is to dress. Most professionals agree that the cutest outfits are just casual clothes: t-shirts and jeans or a sweatshirt and baseball cap. Do not dress your child in elaborate or frilly clothes unless you are instructed to do so. Makeup is strictly out for little girls, as are jewelry and excessive bows. Clothing and accessories should not overwhelm the child but rather enhance his natural charm.

The agent will let you know what will be expected of your child. She may be asked to eat something, to smile on cue, or to do a little dance. The key is to prepare her for what she is going to be called upon to do without over-rehearsing. If there is copy involved, the director may want your child to read it for the first time in his presence.

When To Pull Your Child from the Business

Use your parent's intuition to pick up the signals your child sends. If he is acting up or suddenly not booking jobs, something may be troubling him. If you see unpleasant personality traits beginning to develop, such as an overly developed ego or rudeness, it's time to talk seriously with the child and cut down on auditions, or possibly pull out altogether. No matter how caring or wonderful his agent or manager is, they may not suggest pulling a successful child from the business. Your child's welfare is in your hands.

The Legalities

Each state has different labor laws pertaining to minors. California and a few other states have what are known as the *Coogan Laws*, which provide for a percentage of the child's wages to be held in trust until the child is no longer a minor. In New York a parent or guardian is not restricted with respect to handling a child's earnings.

Most states require work permits. California requires these be renewed annually, while New York requires a new permit for each job. California requires tutors to be on the sets when children are employed, while New York does not.

The union guidelines are in effect nationwide and override state laws when the state regulations are more lenient. Where the state guidelines are more stringent than the union's, the state laws control.

SAG Guidelines

SAG stipulates that interview and fittings be held after school hours, prior to 8:00 P.M., with at least two adults present at all times.

The parent has a right to be present, and within sight and sound of his child at all times.

No child is allowed to work in a situation that poses danger to life or limb. If the child is afraid to work in a situation that she perceives to be dangerous, regardless of how safe it actually is, she is not required to perform.

One person is required to be on the set to coordinate all matters pertaining to the welfare of your child.

A guardian must be at least 18 years old, and have the parents' written permission to act as such.

A parent is entitled to receive the same travel, lodging, and per diem meal allowance as his child receives on out-of-town shoots.

Whenever Federal, state, or local laws require, a qualified child care person, such as an RN, or social worker, must be provided on the set.

The producer must provide a safe and secure area for minors to rest and play.

If a minor is on location, transportation must be provided promptly at the end of his work day.

Work Hours and Rest Time

The work day for minors can begin no earlier than 7:00 A.M. for studio productions, and 6:00 A.M. for location productions. The work day must end by 7:00 P.M. for children under six years of age, and for older children by 8:00 P.M. on school nights and 10:00 P.M. on nights when there is no school the next day.

The maximum work time for minors cannot exceed six hours for minors under the age of six, and eight hours for older children. This does not include mealtime, but does include a mandatory 5-minute per hour break.

A minor must receive a 12-hour rest period between the end of one work day and the commencement of another for the same shoot.

Interview with Fran Miller
Children's/Young Adul Agent
Commercial Division
Abrams Artists, Kronick, Kelly and Lauren
420 Madison Ave.
New York, NY 10017

BEARDSLEY: For which areas does your agency represent children?
MILLER: All areas—commercials, television, feature films, stage, and print.
BEARDSLEY: What ages do you handle?
MILLER: Infants to young adults, which is 18 to 22.
BEARDSLEY: Do you work with babies?
MILLER: I have some babies who I call directly. Meeting babies is very time

consuming. I get lots of pictures. I do prefer to meet the babies and know who I'm sending on auditions, but there's just so much going on that it's a lot easier to have someone else screen them for you. So many times I call managers for babies.

BEARDSLEY: What do you look for in babies?

MILLER: In order to work in the business, they need to be a legal three months, and a lot of times what directors like to do is find a baby who is three months but looks younger, to play a newborn. So for those babies they want little or no hair.

BEARDSLEY: When does a child need a manager?

MILLER: I recommend that a child go to a manager when I've already got three or four kids who are wonderful, the best kids in that age range, and it's hard for me to take on what we call a second layer. Somebody who is just starting out, who may not do well in the beginning until cultivated a little bit more, might do well with a manager.

BEARDSLEY: Which managers do you use?

MILLER: We work with New Talent Management and Shirley Grant Management in New Jersey, and Jeff Mitchell in New York. Renee Courtney also has great kids.

BEARDSLEY: How hard is it to get your child started in the business?

MILLER: Actually, it is not that big a deal. The truth of the matter is that anybody can meet an agent or a manager. If you send a picture into an agent, they will call if they think you're cute. I get hundreds of pictures a week; but if there is somebody we think is special, we'll call him in. There's no secret ingredient to getting in the door. Nothing that somebody else knows is going to work as opposed to what you know. It just takes trying. If you can find out who to contact, that's all you need to know. Send them pictures, if they're interested they'll call you back. That's really what it comes down to. We do take chances with kids, we're always looking for new kids, always looking to bring new kids into the business. There's no secret that somebody else knows that you don't know. A little girl who starred in *Hook*, Amber Scott, is our client. Her parents sent her photo to us about two years ago. The picture came across my desk. I thought she was adorable, called her in, and started working with her commercially. I sent her on a couple of auditions, she met the legit department, they sent her on an audition, she booked a movie. The child had acted in one commercial at this point and booked this movie. Steven Spielberg adores her. She sang a song in the movie—and she's not a singer. She's a six-and-a-half-year-old girl who has never been trained. She has a sweet little girl's voice. The song she sang was nominated for an Oscar, so she's singing at the Academy Awards.

I've got kids on television shows, on Broadway, in movies—my kids work. Anybody's picture can come across my desk and you don't know what can happen. If I hadn't called Amber in, who knows if she would have been in *Hook*? You have to start somewhere. You never know which kid is going to be a star. As an agent, you meet as many as you can and hope that you're finding something good.

BEARDSLEY: What is the best age for getting a child into this business?

MILLER: The kids who have the best chance of working are the little ones, the four-to-five year olds. Those kids don't really have to be capable of much. They don't have to read or have any great acting ability, they just have to be very verbal, very precocious. And there are plenty of jobs out there for the young kids. That's when most of the kids get started. There's stuff for all ages: Four-to-five is very busy; eight-to-ten is very busy. You will find that the older the age range is, the longer the children have been in the business. So, as a child gets older, he competes against kids who have more experience and are more comfortable in the business. That's not always a bad thing; kids can start at any age. Directors do look for fresh faces and new people. If somebody is talented, and has something special, she will work at any age. It's probably that the younger they are the easier it is to start—the less intimidating it is.

BEARDSLEY: Besides the obvious financial advantages, what are the benefits for getting your child into the entertainment business?

MILLER: It makes them comfortable with strangers, it brings them out, it makes them more personable. They become less shy, more independent, more secure. Definitely these kids will go on to become good public speakers, comfortable in different environments. There's some traveling that might be involved. If you're doing a shot outside of your hometown, you travel there. And the money—it's a great way to pay for college educations. Some of these kids wouldn't be able to go to college, or would have to take out student loans. It would be very difficult for their parents to pay for college, and this is a way to pay for it. There are many benefits. Of course there is the excitement. The kid does a commercial one day, he can be in a movie or a series the next day, you never know.

BEARDSLEY: Psychologically speaking, how do you keep your child from turning into a little adult?

MILLER: It depends on the adults and how they insulate the children, and that's the honest truth. How you explain things to them, how *you* handle the pressure. For the most part, kids can deal with this. Kids have fun going on auditions. It's how much pressure their parents put on them that hurts them. Parents start saying such things as, "If you don't book something soon, we're going to have to stop, we can't afford going to the city three times a week." Or, "You know, if you book this commercial you can go to a private school like you want to next year." If the kid's got a big test, it's OK to say to the agent or the manager, "Sorry he's got a test, he really needs to study tonight." If the kid wants to play soccer, let her play soccer. If it's going to work, it's going to work. I have kids who know they're supporting their family. So, maybe it's a single parent family, and in order for the kid to do this Mom can't work. They know that. It's a lot of pressure to put on a 10-year-old girl. For the most part, the parents get more upset, more frustrated over this than the kids do. I will often hear a parent say, "I didn't book that did I," and I'll say "No, *he* didn't book that." Parents personalize it, a lot of parents want to do this themselves. This is something they never had a chance to do. They want to be celebrities; they want to be successful, so they transfer that onto the child.

BEARDSLEY: How do you know when you've gone too far?

MILLER: By watching your child. Knowing your child. Knowing that if she's acting up, there's a reason. Is she starting to feel sick? Does she not want to go to school? There are lots of symptoms. They'll give you the signs. Maybe your child's not booking anymore. Maybe she's not going in there and doing the things that she's supposed to be doing because she doesn't want to be a part of this.

   Listen to your child. If he's constantly saying, "Can I go to my friend's house after school?" he doesn't enjoy this. If he's dreading this, if he doesn't feel well, doesn't want to do this, doesn't want to do that, he's sending a signal. Take your cues from your kids. As an agent or a manager, if anything got back to me, or if something happened on an audition or on a set that might indicate that the child is not happy, I would definitely report it to the parents.

BEARDSLEY: What is your definition of a stage parent?

MILLER: Somebody who is so completely obsessed with the child's career that it has become their life, their career. They think their child is the best, and they want him to do everything he is capable of doing—*everything*. There are a lot of them out there, too. It's no joke, and, yes, it's a stereotype, but they do exist; and anybody who takes their child on an audition is going to run into these types. They want to know who everybody is and who's booking which jobs. There are parents who know of every single job that's out there, every single job that books, who books what, they know that a job gets booked before I know that a job gets booked. They'll call me up and say "Did you hear that so-and-so did this job and so-and-so is doing this?" It's ridiculous. There's one parent we call "Commercial Patrol." She knows every job, who's working on it, and how it's running. It's sick. Her television is on 24 hours a day.

BEARDSLEY: What are the kids of such parents like?

MILLER: They're not normal children. They've also become obsessed with the jobs and auditions and booking and going out. It's really scary, because the truth of the matter is that a good agent is not going to send you on every job. There are certain things you're right for; others you are not right for. Because of that you want an agent who is selective, who only sends you on things that you're going to book, who is not going to waste your time. As an agent, I've got to give my casting directors the best people for the job. I'm not going to send someone who's going to waste their time and my time. Some people want to go on every single job. The kid may be eight years old and white, and the mother finds out that a job's going on and wants to know why he wasn't sent on it, and it was for a 10-year-old black boy. As an agent, there are just certain guidelines you go by. Parents think that if they're not on it they can't book it. So they've got to be on everything. It's frightening. It's a nightmare.

BEARDSLEY: Generally speaking, how demanding is this business on the kids?

MILLER: It's tough. You could have one audition at 3:30 and another at 5:45. You're not getting home until 7:00 at night. School work suffers. These kids are in the city in the afternoon which is normally when kids are doing their homework. A lot of the kids who take the train in do their homework on the train. If there's an audition late in the city, they don't get home until 8:00 at night. Some kids are too tired to go to school the next day. Some kids don't go to school at all, they're tutored at home.

BEARDSLEY: Who do you recommend for photographs?

MILLER: I personally think Jean Poli on East 31st Street in New York is the best for kids. Bob Cass is also very good, also located downtown. And Susanna Gold, who's located in Connecticut now, comes into New York for some scheduled sessions.

8

□ □ □
□ □ □
□ □ □

Voiceovers

The voiceover area may be the most desirable and most competitive of the commercial industry. It is desirable because of the anonymity factor—the ability to work and yet not be seen on screen—the tremendous money that can be made with a proportionately smaller amount of work—successful artists make upwards of $350-500,000 yearly, and perhaps best of all, the fact that appearance doesn't count the way it does on-camera. You can have crooked teeth or not be a commercial type and still be successful. On the down side, the anonymity and money factors make voiceovers attractive to all levels of talent, so you'll be competing with the most successful people on even the smallest jobs. The business is seen as being a closed-door club or clique, open only to a few successful actors who control all the jobs because the most successful voiceover artists of today, and in some instances stars and celebrities, compete against the newest kids on the block. While in on-camera spots new faces may be desirable to typify the slice-of-life aspect the advertisers are trying to portray, in voiceovers it's usually the tried and true that wins the job and eventually sells the product.

Who Works

Neal Altman, head of Abrams Artists' East Coast office and one of the founder of their voiceover division, describes the types of performers who are successful in voiceovers today. The first type, perhaps the most highly compensated, he says has an

extremely distinctive sound and has carved out a niche for himself. While there may be one or two others in the business who can come close to approximating what his voice sounds like, it's a signature sound. Interestingly enough, he may audition less than most people and win a fewer number of jobs, but his voice is so distinct that agency people will hear it on the air and then write with that voice in mind. Precisely because his voice is so distinct and there are so few people who can do what he does, this type of talent can command a premium price-wise.

The second type is the announcer who has been in the business for years and does extremely well. According to Altman,

He probably auditions a lot and works a lot, maybe several hours a day in the studio, doing industrials, sales films, radio commercials, and slide shows. In fact, he probably narrated half of the biology and history films you saw in high school. This type makes it in volume: he's competent, he reads, he's pleasant to listen to, he's not so much in love with his own voice that his voice will get in the way of the copy, he's just a real good announcer.

Type number three is the great interpreter of copy. This artist

interprets copy differently from most because of his acting ability. He has a certain mind set, so that he picks up a certain piece of copy and reads it in a way that's not predictable. While his vocal quality may not be the greatest— he may not have a very mellifluous sound, or he has a heavy, gravelly, or gritty sound, or a driving sound—the way he reads can overcome that and he is chosen because of that. The competition is very fierce in the voiceover area because there are a lot of people who are very good at what they do. But the wild card in that deck is that the writer, director, art director, or copywriter often don't know what they want until they hear it. They can hear 30 men or women and not make a decision. They can hear 30 more and still not make a decision, but somebody could come in, and boom, *that's what they have on their minds. So it's an intangible thing, and somebody who can understand what the director wants and make the adjustment and lay it down and compete successfully—that ability separates the journeymen from the really successful ones.*

Trends and Types

The relatively new trend of hiring celebrities doesn't make much sense to those whose business it is to supply voiceover talent. Ad agencies spend millions of dollars on household names whose voices don't carry the same level of recognition as their faces carry. While most people can easily recognize the distinctive voices of Sally Kellerman or Burgess Meredith, these same people would probably be surprised to learn that Donald Sutherland, Gene Hackman, and Michael Douglas can be heard on, respectively, Prudential Insurance, United Airlines, and Infiniti car commercials. While the ad agency may feel that a Michael Douglas brings something to a commercial that no other actor could, the significant sums of money that it costs them to hire him may help the novelty to wear off quickly.

Another trend is for writers at ad agencies to put their own voices on spots, because they feel that they know best what has to be done, and that they are not going to sound like any other announcer on the market. Some of these spots have been very successful, and in fact talents such as Hal Riney of Hal Riney & Partners ad agency, who does the Alamo Rent-a-Car spots, has become a much requested prototype.

What ad agencies seem to be seeking these days are "non-announcing", "hip," "unusual," "with an edge," or "unpredictable" type voices. Interpreted, this may mean gravelly, raspy, very deep, with a little "texture" or "bite." They sometimes ask for middle-of-the-road announcers, who can deliver tag lines or copy at the end of the spot. They may be casting for a spot that is dealing with something uncomfortable to a lot of people, income taxes for instance, and need a voice with a warm and reassuring quality. They may have a dandruff spot that needs someone to "drive it home." They may be casting for a new toy and want it to come across as the newest, most exciting thing to ever hit the toy market. They will ask for someone to "punch it out" like a Marine sergeant, someone who has a lot of guts and drive in his voice. Also requested are people who can read things in a dry, ironic way, such as the late Lester Rawlins, the voice of Dunkin Donuts and NYNEX Yellow Pages, or Burgess Meredith, who can be heard on a myriad of different spots.

Prototypes of specific people are also in demand. "Tammy Grimes-ish," "Sally Kellerman-ish," "Hal Riney, James Coburn, or Martin Sheen type" sounds are frequent requests. If there is a successful campaign on the market featuring a certain celebrity, that person will most certainly become a requested prototype.

Women and Minorities

Unfortunately, there is less work for women and minorities in voiceovers, even more disproportionately so than in either on-camera commercials, television, or motion pictures. Women have to be that much more talented, distinctive, and determined to find a place in this market. You can prove this to yourself with an at-home experiment. Take one hour of prime-time television for five consecutive days, and list every time you hear a male voice versus every time you hear a female voice. Or listen to the same ten minutes of radio every day for five days and write down every time you hear a man announce or do a character versus every time you hear a woman. You would probably find that nearly three-fourths of the voices are male.

So, too, with minorities. An odd racial bias exists. Oftentimes, when copy is geared specifically towards the Black or Hispanic market, producers look for a certain sound. They may say they want a "black sound" or that they don't want the announcer to sound too "street." "Street," "urban," and "black" are commonly used descriptions. The irony is that many minority announcers can't compete on this level because they don't match the sound the ad agency people are looking for. Having fewer opportunities to begin with, they have even less of a shot at voiceovers. On general calls, good actors who happen to be Black or Hispanic might or might not be seen. If they have no accent whatsoever, they have a decent shot at getting in on the call. There are minority announcers who do very well, not because they are Black, Hispanic, or Asian, but merely because they are good announcers.

Who Should Try and How

Distinctive voices have a better shot than middle of the road voices. Most important for those who want to do voiceovers is to recognize that they have to fill a certain niche. Although most actors think they can do most things, it's a matter of what you are going to get *hired* to do. When you compete, you have to have something to sell that will make you more desirable than somebody else.

There are people who have great voices who hear all the time, "Oh! you should do voiceovers." But they don't know how to get in there and compete successfully, so they don't get hired. The technique for being desirable as a voiceover talent is something that can't be taught; it must be acquired. It's a classic *Catch-22* situation: You can't get people to send you out unless you have good material on tape, and you can't get good material on tape until you have competed successfully.

Demo Reels

In the voiceover area there are two distinct types of demo reels. One is produced in a voice studio with an engineer, using copy either cut out of a magazine or provided by the studio. There are many recording studios that make a nice living preparing demo tapes. These are known as *B-level tapes*.

A-level tapes are the resume reels, which consist of actual commercials the performer has done. You can run two or more consecutively on a tape in their entirety, or create a medley of edited portions. A tape prepared from existing spots is far ahead of a tape made in a recording studio, for the simple reason that the A reel consists of spots that someone has paid the artist to do.

B tapes are useful, however, if only to see if people will listen to it to determine an actor's potential. These tapes will probably not get an actor work for a couple of reasons: If somebody is making a decision for a job based on listening to a handful of tapes, and included are resume reels of people who have been doing this for a while , their tapes will shine above a produced tape. Secondly, there is a recognition factor in a resume reel. When you hear a spot on the reel of a working voiceover talent, you may say to yourself, "Oh I recognize that music, I remember that talent." Produced tapes utilize a type of canned music that the engineer has put on there. In fact the engineer, who has done hundreds of these tapes, most likely has used the same music on the same copy on the same spots, resulting in a flat, uninspired sound.

People who are new to this business sometimes think they can do anything. They may seek to try a "romantic" read, a "hard sell" read, a "hip" read, or a "laid back" read on their demo tapes, when in fact they've never been hired to do any of these. Making a B tape amounts to a shot in the dark. In order to work a B tape must be good. Otherwise it could be the kiss of death. An agent gets 20 unsolicited tapes a week, over a thousand a year, so in order for yours to stand out it must be targeted well.

If you have a very distinctive or unusual sound, either raspy, gritty, high-pitched, or husky, it may pay to produce a rough demo. If you do character or cartoon-type voices, include them. The trade papers are filled with ads for recording studios that produce reels; one is just as good as another. It is not necessary to pay top dollar to get your voice on tape: you may spend hundreds of dollars trying to get an air quality tape that still won't equal the quality of a resume reel or represent what you may be hired to do. This tape should be produced only for the purpose of letting people listen to it to see what they think of your potential, not mass produced and sent to copywriters and casting directors to be considered for jobs.

Voice Classes

A voiceover class is a good investment because it gives you a place to go every week and work at what you are trying to do. The danger is that these classes sometimes take away from what makes a performer unique or distinctive and turns him into a slick, familiar kind of announcer. If a student observes a certain style, and learns to work on it so that every adjective is punched, to go "up" at the end of the first line, and "down" at the end of the last line, and alters his inflection the same way every time he reads, then he sounds robotic, as opposed to interpreting and delivering copy in a way that might make him stand out better. Some of the classes are helpful in assembling a tape, and at the end of the class you may have your first demo reel. While this tape won't necessarily get you a job, it will probably be enough to have people listen to for their opinions.

Breaking In

If you are confident in your skills and you've had an engineer produce a demo reel for you, you now must be perseverant in trying to get in to see agents and casting directors and recording studios—places where people can give you some feedback. You need to know if they think you have potential and whether they can find an audience or opportunities for you. There is a great deal of rejection associated with this, for most times people are just too busy to listen. Your best bet is to network to a point where you can get an introduction to a voiceover agent, casting director, recording studio engineer who can get you a meeting with an agent, or a copywriter who can request you on auditions and introduce you to a casting director.

If you have a distinctive sound and the flexibility to move to a smaller market you can try to pick up work there. Actors who do regional theater can call up local recording studios and advertising agencies and say, "I know how to read commercial copy, I'm competent at it, I'm in town and available for a couple of months. Can I come and see you? If you need somebody to do radio commercials, while I'm here, I'm available." Those people who are able to work in a smaller market and then go to New York or Los Angeles with a resume reel are one step ahead of the game.

Breaking into the voiceover area is not so different from other areas. Perseverance, knowing the right people to talk to, finding people who are successful in one form or another to help you, will all help you in your quest to get you where you want to go.

Compensation

Even though union minimum is less for voiceovers, voiceover artists have the ability to earn more, while at the same time experience less exposure. Most successful voiceover people work for scale. An on-camera actor may do a commercial for Maxwell House at scale, and for a day's work may earn $10,000 over the course of a year. The announcer who does the tag line on that same commercial may also have his tag line on every Maxwell House commercial in that particular campaign or over that period of time. It is not inconceivable for that announcer to have his voice on eight or ten commercials for Maxwell House, and to earn upwards of $100,000 at scale for just one line— one hour's work in the recording studio.

The Audition

Depending on the casting director, you may be told they are looking for a specific sound and offered some direction, or they may say that they want something "clever," "why not try it," and then direct you based on what you do. You usually get two takes, one to do what they ask you to do, and one to see if they can direct you a little differently from that. Those who are nervous and freeze a little may get only one take and a "thank you very much" from the casting director.

There is no rule of thumb in terms of how auditions are run. You may walk into a room with a little mike and three or four people standing around you, or you may walk into a fairly modern advertising agency equipped with a sound booth, an engineer, and a mixing board, or anything in between.

In general, there is less improvisation on a voiceover audition than on-camera, and there is less to worry about in terms of what you look like and what you are wearing. While you shouldn't go in after having just run six miles, if you are wearing jeans and a t-shirt and the next person comes in wearing a suit, one doesn't necessarily have the edge over the other.

Interview with Carol Hanzel
Casting Director
Hanzel & Stark Casting
39 W. 19 St., 12th floor
New York, NY 10011

BEARDSLEY: What kind of casting does Hanzel & Stark do?

HANZEL: We basically cast commercials, both on-camera and voiceover. That is about 90% of the work. The other 10% is divided between theatre, industrials, and television.

BEARDSLEY: Do you personally work in one specific area?

HANZEL: We both do just about everything. My partner also specializes in casting for the Hispanic market. She is fluent in Spanish and does a lot of on-camera and voiceover herself.

BEARDSLEY: Do you feel that your experience as a performer and an agent has helped you as a casting director?

HANZEL: Well, number one, I am empathetic towards the actors. I know what they are going through. I know what it is like to stand up there and sell yourself, and I certainly know from an agent's point of view how difficult it is to get your people started. We as casting directors cannot always see everybody the agents feel we should see. There is a limit on the number of people who can be seen for each project. That limitation is put on us by our client.

BEARDSLEY: What kind of limitations do they put on you?

HANZEL: For example, they say, "You have one day to cast eight roles." That does not give you a great deal of leeway for mistakes or for trying a lot of new people. You can try several, but you cannot devote a session to trying to uncover a wonderful new talent.

BEARDSLEY: How many people do you generally audition for a single job?

HANZEL: I am very heavily into the voiceover field. I do a lot of them and I work very quickly, so for voiceover, if I were looking for one person and doing a full day of casting, I could bring in 45 people.

BEARDSLEY: How many of those might be newcomers?

HANZEL: Maybe five. Also, we get a lot of unsolicited tapes in the mail.

BEARDSLEY: Do you listen to those?

HANZEL: Every one of them.

BEARDSLEY: What do you think about the quality of those and the chances of getting in to see a casting director of your caliber that way?

HANZEL: It is extremely competitive. Of more than 50 tapes from both men and women that I have listened to in the last two months, I have kept only three. The rest have either been very amateurish or well produced but not different enough voice quality-wise or delivery-wise for me to pick up the phone and say, "I'd like to meet you."

BEARDSLEY: Do you think that legit actors make the best voiceover artists?

HANZEL: Not necessarily. I think if they happen to have a very interesting voice, as well as acting skills, they are wonderful because they approach the copy from an actor's point of view rather than just as a straight announcer or like someone who has no acting background. They can color a reading. They can give it a different slant, but if they do not have an interesting voice, then no matter how good an actor they are, it is not really going to cut it. For instance, Martin Sheen was the perfect person to break through. He had a very interesting, distinguishable voice. It was different from what was out there and he had great interpretation. So, boom, within a month of his saying, "You know, I would like to do voiceovers," the guy was really booked.

BEARDSLEY: What does a voiceover audition entail?

HANZEL: Let's say I am looking for one special person for a TV voiceover. It starts with the agency who sends me the script, and we talk about what they *think* they are looking for. I say "think" because what they start out wanting and what they eventually choose are two different things. But, that is part of the process. So I listen, they talk, and then I think about my wish list, who I would like to bring in. Once I have my wish list, I call agents and give them the breakdown, tell them what I am looking for, and ask if they have any thoughts other than the people I have requested to bring in.

BEARDSLEY: Do you usually get your wish list?

HANZEL: Pretty much, unless they are out of town. Being out of town is not a big deal because we can always find a studio someplace somewhere where someone can walk into the studio and put the audition down on tape and FedEx it.

BEARDSLEY: Which agents do you work with consistently?

HANZEL: If I were just talking about adults, for voice, on a single job I probably work with five to eight agencies. These agencies have tremendously strong voice clients, so I work with them all the time: SEM&M, Abrams, J. Michael Bloom, Cunningham, Buchwald, LW2. You may or may not have heard of LW2. They spun off from Susan Smith & Associates and have a nice, small, boutique-style agency. Not a huge client list but pretty much a winner client list. Another agency like that is STE. We do a lot of work with them, too. William Morris, of course; I have to call over there for certain people. I can't not do a session without certain of their people. I am leaving out TRH because they are not my first call, but they certainly are on my list.

BEARDSLEY: How important are the agent reels? Are they useful to you?

HANZEL: Yes, but they are only useful to me if I have to reinforce a thought to a client or if we couldn't get that person in. I'll say to them, "Why don't you listen to this and let me know if you want me to pursue it." So it is important for the agency to have a reel, very important. In terms of how often I go to an agency reel, not that often. I do more of that when I am looking for out-of-town talent such as the California contingent. I have all the California tapes and there are a lot of jobs that originate here in New York; but, as I said before, since people are just a phone away or FedEx away, we don't rule out California talent if it's a big project, not a local radio or a small TV run. But a major campaign, let's say, Jeep Cherokee, which was a big, big campaign I happened to cast recently, means a nice bit of change to somebody. So for something like that I will talk to them about certain people in California, and put them down on tape. If not for Jeep, then AT&T, or Hardees or anything like that. For any of the big campaigns you would also include L.A. talent and even Chicago talent.

BEARDSLEY: Is there a problem in selling the talent that you know would be the best choice?

HANZEL: That's always either a joyous thing or a disappointment. It is one of the two because you do a session, you knock yourself out, you bring in great people, and the client will turn around and say, "You know, there was nobody on the tape we liked." I will be stunned then, because for these types of jobs I will have brought in top, top people, and any one out of 20 could do the job; but for some reason the client is on a different wavelength. So, that's always a disappointment, but the joy is when you say, "This guy or this person was so great, or this group was so terrific . . . you have to give them special attention. You have to go back and listen," and they come back and say, "They *were* great. Book them." It is always fun when you are in agreement.

BEARDSLEY: So what do you do in the first scenario when they seem dissatisfied? Do you then bring in second rate talent?

HANZEL: In my mind, I think they missed the whole point. They missed the A List that we just did. But, again, it is such a subjective business—it is not my product, or my storyboard, it wasn't my idea, so I'm there to, like, park the car. I am there to serve, to bring in the best people I can bring in who are free to come in on this project. If I have missed anyone, I am sorry, it happens. But, we are pretty damn thorough. We give as much attention to the smallest client, who may have one account, as we give to the largest ad agencies in the world. That's how we work. I am very picky.

BEARDSLEY: What are the mistakes actors make?

HANZEL: Oh, I think coming in with an attitude. I can appreciate it, because I was an actress. I can look back and think, "Boy, were you stupid then. If you had known then what you know now, unbelievable." Hindsight is a wonderful thing. Coming in with an attitude or coming in and not listening. My partner and I are there to make the actors look good and also make ourselves look good. We want everyone to look good. We want the agency to look good so that when the client, the ultimate, last person, sees everything, he says, "Great, this is a really good job!" When actors come in and don't listen and think that we are not in their corner, that's a mistake. Why wouldn't we be?

BEARDSLEY: Which voice types are most in demand?

HANZEL: The sort of bad boy type. Bad boy meaning like the Alec Baldwin type of voice—the kind of guy who would spell *trouble*. Those are very much in demand, and the age range on those voices would be late-20s to mid-30s. Smoky, sexy, with a texture—and the same thing for women. The Demi Moore type of voice is very much in demand. We call it the damaged voice. What is not in vogue anymore are the straight announcers.

BEARDSLEY: What's happened to the guys who are used to making a living at straight announcing?

HANZEL: They're looking around for work. A lot of them had big accounts all through the 60s and 70s that they could count on for regular income. Just like fashions change, so does the style of voiceovers. They are a little bit on the outs right now. But it is cyclical. That style may come back. It is hard for them because they were used to making a very good living by doing a certain thing, and now that is not in vogue. They are getting older too, and, as we know, advertising is a very youth-oriented business. Extremely so. That's the reality of it.

BEARDSLEY: Do you have any advice or direction for the beginning performer?

HANZEL: I think the best advice I could give someone is don't invest in an expensively produced tape right off the bat. Nine times out of ten it's going to get you nowhere. I think probably the thing to do is go into a small studio and put down as clean a representation of yourself as you can. Unless you do incredibly fantastic character voices, don't bother doing different dialects. People make the mistake of doing a couple spots as themselves and then doing an Italian accent, an English accent. It doesn't do them any good. The thing that you want to establish is an overall personality on a tape. Whether it's that you are just a wonderful character person with great comedic timing, or a very sultry cosmetic-voiced person, you have to establish some sort of continuity. You can vary in the spots you do, but it is very confusing when a tape is almost schizophrenic in nature. We know it if you're an actor, and if you're a pretty good actor you can do a lot of things, but you are trying to get someone's attention, and, unfortunately, this is a business that pigeonholes voices as well as faces. You have to, otherwise it would be impossible to try to narrow down a casting idea. Let's say that if I am a female voiceover artist who happens to have a very clean bright sound, not a sultry sound, I am not going to try to present myself as doing many characters or other aspects that will only confuse the tape. I also think it is important on a tape to have a spot where you work with someone else so that it's not just you, but offers a back and forth. If you are a woman, get a man to read a spot with you and vice-versa. That is important. I know that on one of our earlier [casting promo reel] tapes, which was an incredibly strong tape with mostly network TV voiceover spots on it, someone said to me, "Gee, I would have liked to hear some of your radio stuff where you have multiple voices," and it made me think. Our current reel is mostly radio. It's a lot of stuff with different people on it so that they can hear the range of people we go for. Some of it's serious—PSAs for Drug Free America—and others are as wonderfully silly as Laughing Cow. When some people put together a tape, they think they are so versatile that they want to show every side of themselves. Unless you really are established at that point, it gets very confusing to the people who listen. Or unless your forte is that you do character voices.

Here's a story of an agent I worked with who does a lot of work with stand-up comics, for on-camera for television and so on. She met a stand-up comic, saw his act, and noted that he had an interesting, distinct voice. She asked if he had ever done voiceovers and he said he had never even thought about it. Well, she said he should come into the office and talk about it. The guy went in, and on his first audition, he booked. She immediately signed him. Now he hasn't broken yet, hasn't hit the top, but he is about to, so I hesitate to mention his name. This was one of those situations where the agent recognized that his was a unique voice, and he was a comic—comics generally have great timing—and he just hit, he connected. I can't say that he knows what he is doing, but he doesn't have to. He is just so natural. So that is always fun. I had him in on maybe his second audition and said to my partner, "This guy is going to be hot." You just know it, you hear it, all of a sudden you go, "That's something different, I haven't heard that before."

BEARDSLEY: How often do you hear one of those?

HANZEL: Not often, not often at all. So when you do hear it, you really do sit up and say, "Whoa!" You are also hot if you reflect what is going on in the entertainment industry, if you happen to sound like someone else who happens to be hot. I don't mean a rip-off, but remind people of that person. Then you immediately stand a chance of booking jobs because there is a familiarity concept there. They say, "Gee, you sound so much like . . . Sally Kellerman." They [the clients] say they want Sally, but can't really afford her, so they ask, "Who do you have who sounds like her?" It is always hard when you do casting like that. You don't want to rip off Sally's voice, but you are looking for a similar attitude. Before you know it, you are almost doing line readings that sound just like her.

It is hard for people to be objective about their own talent, and usually family and friends are not the people to discourage you. But if you really do think that you have something salable, then go ahead, pursue it, and if you get turned down too much, I think then you have to say, maybe it isn't as salable as I think. But I have seen it happen. I have personally been there when it happened, when a tape has come across my desk and I have called somebody in to tell them they are very good, and need whatever. I fix them up with agents because we all know that you need an agent. I have no problem doing that because I probably get to see the talent before the agents do—we get to see so many more people, whether it be through shows, recommendations, or tapes. We have hooked up a lot of people that way. It is not entirely out of the question. It is difficult though. Know yourself, *know yourself.* You can make a very nice living if you click in voiceovers. I don't know for how long you may make that. You may be so hot that you outdistance yourself—where you are so exposed that nobody wants to hear you for a while. That tends not to be quite the case as much in voiceover as it is for on-camera. In voiceovers there is one fewer sense than you have for on-camera where you see *and* hear a person, as opposed to when you just hear that person.

Carol Hanzel and her partner Elsie Stark have been in the business for a combined total of 20 years, casting upwards of 2,000 voice projects in that time.

9

□ □ □
□ □ □
□ □ □

Industrials

Recorded Industrials

Recorded industrials, as defined by AFTRA, are nonbroadcast recorded material used for instruction or education, sales promotion, amusement or entertainment at meetings, conventions, points-of-sale, in-plant, public displays, churches, classrooms, seminars, or any other site or location of traffic, congregation, or convening, or transmitted by phonecasting, laser or other photo transmission equipment. For actors, this most typically includes training or management development films produced by corporations for their own in-house instruction, or generically produced films packaged by a single company that sells and distributes the films to various corporations. An *industrial program* is nonbroadcast recorded material produced for one client on a single subject and released as a package.

Acting Types Used

The types most in demand are straight spokespeople, men more often than women, to narrate the films. Comedic actors, or types with an edge, women and minorities are becoming more popular as companies begin to accept nontraditional types as authority figures. Industrials seem to be the entertainment world's last bastion of evolving types: Film or television programming usually does the groundbreaking by imitating life; followed by commercials, commercial print, and finally, industrials, as each in turn adopts or mimics the trends.

Also cast are the role players: people to play the parts of CEOs, office employees, line workers, the families of employees, and so on. Women and minorities are used more frequently in this area, but still somewhat stereotypically. Children are used when the families of employees are part of the scenario.

Occasionally taped industrials employ singers and dancers, but as this greatly increases the cost of production, their use is generally restricted to live industrials.

Compensation

As residuals are not part of the pay structure of industrials, the compensation can be significantly lower than it is for commercial work. But since there are no conflicts, actors are free to do as many industrials as they are able to book. Many actors have recurring clients that use them over and over again, enabling them to command increasingly higher salaries. Actors employed as a given company's spokesperson who have a high profile in television, film, or theatre command impressive day rates.

Union Wage Guidelines
AFTRA breaks down the employment of actors into

- day player
- three day performer
- weekly performer
- on-camera narrator/spokesperson
- extra

On-camera narrators or spokespersons are performers whose primary function is to explain, demonstrate, instruct, or promote, substantially in monologue. No more than two on-camera narrator/spokespersons can be employed in any one program.

Taped industrials fall into two categories:

Category I: Programs that are designed to train, inform, promote a product, or perform a public relations function, and that are exhibited in classrooms, museums, libraries, or other places where no admission is charged. Included are closed-circuit television transmissions such as direct broadcasts by satellite and teleconference. Also in this category are sales programs designed to promote products or services of the sponsor, but which may be shown on a restricted basis only.

AFTRA cites as examples: a program designed to inform sales representatives of the features of automobiles sold in their dealership, or a program promoting fire prevention which does not attempt to sell a particular company's fire insurance policies.

Category II: Programs intended for unrestricted exhibition to the general public. The program must be designed primarily to sell specific products or services to the consuming public at locations where they are being sold or at public places such as coliseums, railroad stations, air/bus terminals, or shopping centers. Also included are those programs supplied free of charge to customers as a premium or inducement to purchase specific goods or services. A 5-year use limitation is imposed.

AFTRA cites as examples: a program outlining the selling features of an automobile which is available to all consumers entering automobile dealerships, a program promoting fire prevention which sells the benefits or a particular company's fire insurance policies and is exhibited in a shopping mall, or a video cassette explaining how to build a recreation room, provided free of charge to anyone who buys a power drill.

Preparation for Finding Work

The tools necessary for finding work in industrials vary slightly from the commercial area.

Pictures

Headshots used for taped industrial work should be straightforward and should look like you. They should not be overly glamorous or made up, but should present you as a management or blue-collar type. If you find yourself being considered for spokesperson roles, invest in a straight spokes shot. Otherwise, your regular commercial shot will do.

Resume

Create a resume targeted for industrials, which highlights your acting experiences and covers your industrial work, including types of roles. Although industrials do not hold conflicts, veteran performers are better served by listing the production companies or *types* of corporations they have worked for, instead of naming company names. For instance, Sprint might be leery of hiring an actor as a spokesperson whose resume lists a previous job with AT&T.

Don Snell, marketing director for Corporate Productions, Inc. and veteran of more than four dozen industrials whose full-length interview is included in Chapter 3, advises actors to list their most important television, film, and theatre credits on the top half of a resume so that they can see your acting experience first. On the second half indicate the industrials you have done by name or producer and types of roles. On the bottom right 1/8th of the page, he suggests leaving a 2-inch column blank for notes. He promises this format will be well-received by producers and casting directors. Adds Snell, "Include your contact numbers on top, and write in your age range in pencil."

Video Reel

As for broadcast commercials, a video reel is only worthwhile when you have professionally produced material to include on it. When you do industrial work, arrange with the producer to receive a 3/4-inch copy for yourself while still on the set. Do not wait until after the fact to try to get a copy. After you have a few roles to your credit, put together a demo reel. Label each copy with your name, address, and phone number on the top of the tape as well as on the spine, which is visible when the tape is stored.

Making Contacts

Backstage prints a list of industrial producers; AFTRA and SAG provide lists, too. You do not need an agent to approach these sources. Industrial producers, unwilling or unable to use an independent casting director often do the casting themselves. For this reason you should make sure that all the industrial producers have your picture and resume on file. A simple cover letter letting them know why they should avail themselves of your services, along with your picture and resume, is all that is required. A follow-up phone call to find out if they would like a copy of your video reel, even to out-of-state producers, is well worth the investment. Many producers can and do call actors directly from their files.

Mail your industrial picture and resume to local casting directors as well, with a note letting them know you are available for this type of work. Although many call agents, some casting directors contact talent directly for industrial work. It is best to have all ends covered.

When To Use an Agent

It is not a bad idea to refer offers of employment to an agent to see if she can better the deal for you. If you are reluctant to hand over your hard-won money, remember that one job usually leads to another. Referring negotiations to an agent will show that you are a working actor and will keep you in that agent's mind for other projects. Also, very few actors are able to negotiate favorable conditions for themselves. Most do better to keep their creative and business activities separate and let a third party do the dirty work. Even if there is little leeway with salary negotiation, a reputable agent can ensure that the job is legit and done according to union guidelines.

Non-Union Work

If you are not yet a member of the union and are empoloyed for a nonunion job, make sure that the producer or casting director knows *up front* that you will require payment in cash on the set. Do not let this slide. This is not an unreasonable request and is one way of ensuring payment.

Live Industrials

Many corporations make use of the live industrial, or *business theatre* as Equity denotes it, at their annual sales meetings, product launches, or other events. The use of live entertainment serves not only as a well received perk, but is useful for educational and motivational purposes as well. According to Cary Chevat, Director of Sales for Kevin Biles Design, a Los Angeles-based industrial production company, live entertainment helps make these events, which are attended by the top sales people and managers, something to look forward to. They are often held in Hawaii, the Carribean, on cruise ships, or some other resort destination. Said Chevat, "IBM started doing them about twenty years ago, and now most major corporations have them as well."

Most in demand are Broadway caliber dancers who sing, although singers who dance may be requested as well. Live industrials consist of small casts, usually five or six dancers or singers, and assorted speakers who either work for the company or are employed especially for the event. There is generally a headline talent: Chevat cited recent examples of Ben Vereen, James Garner, Kenny Rogers, and the Smothers Brothers. Budgets run from half-a-million dollars to upwards of three and four million. Although Biles Design has been lucky enough to escape the crunch, according to Chevat, the industry as a whole has been hard hit by the recession. Translated, "This means budgets of three-quarters of a million, instead of a million." Unfortunately, the live talent is usually the first to go.

Union Guidelines

Live industrials can run from a couple of days to several months. They may occur virtually anywhere in the world. They involve varying degrees of difficulty, from simple performances with light rehearsal schedules, to full blown productions requiring intense and lengthy rehearsals. For this reason, Actor's Equity Association goes into great detail to delineate the working terms and conditions for this type of employment. What follows is a synopsis of Equity's current Agreement and Rules Governing Employment in Business Theatre. For a more complete description of rates, conditions, and contingencies, be sure to obtain a copy of the guidelines for yourself.

Salary, Benefits and Other Compensation

Effective as of 5/31/93, an actor employed for a period of two weeks or more is entitled to a scale payment of $904 per week. Minimum compensation for actors employed for a period of seven calendar days is $1132. If employed on a daily basis, the rate for the first day is $376, with payments of $188 each day thereafter.

The performer is entitled to receive full contractual salary from the first day of rehearsal up to and including the day on which he returns to point of organization or place of engagement.

Overtime is computed on a basis of 1/32 of the performer's weekly contractual salary per hour, and if employed on a daily basis, 1/6 of the first day rate per hour.

A per diem of $60, paid in advance of each week, is due the performer when the place of performance or rehearsal is not within reasonable daily commuting distance.

If a contract of employment specifies a term of more than two weeks or of one week or less, that term is the guaranteed period of employment and salary. If the contract specifies a term of two weeks or does not specify a term, the minimum guarantee of employment and salary shall be two weeks.

If the closing week of employment is less than seven calendar days, the performer's salary is prorated on the basis of 1/6 of the weekly salary for each day or fraction thereof, or 1/32 of the weekly salary for each hour or fraction worked thereof, whichever is greater.

All additional monies earned by the performer are to be paid no later than 12 business days following the week in which they were earned, or an additional 10% shall be paid for each week, or part thereof, that the payment is late.

An actor may accumulate valid sick leave with salary at the rate of one day a month, up to a maximum of six days, for each production.

An agreed-upon weekly rental rate is to be paid the performer on a monthly basis or at the end of engagement for clothes and/or costume rental. This rate cannot be less than:

formal ensemble (tuxedo, tails, evening gown)	$40
business (suit, dress, sport coat and slacks)	$20
sportswear (shirt and pants, sports apparel)	$15
outerwear (topcoats, including scarf, gloves, hat)	$10
dance shoes	$10

The performer must receive reimbursement for the cost of cleaning. However, the producer and actor may agree in advance to the amount to be reimbursed.

Taping fees are due the performer under certain circumstances. There is no fee for noncommercial, nonbroadcast archival filming or videotaping, provided it is made during a regularly scheduled rehearsal or performance. If there is a special call for such a film or videotape, the performers are to receive not less than the AFTRA or SAG on-camera principal rate for industrial films, in addition to the Equity contractual salary. Use of such videotape or film is expressly restricted to the producer's or client's in-house purposes only.

Sound recordings not used commercially or publicly, but to be distributed as souvenir records to the audience or invitees, may be made provided the performer is compensated at the rate of not less than an additional 1/6 of a week's contractual salary, or the minimum AFTRA salary, whichever is larger, for each day or part thereof employed in making the recording.

Radio broadcasts entitle the performer to a minimum of an additional 1/6 of a week's contractual salary or the applicable AFTRA rate, whichever is higher. This rule applies whether the broadcast is live or prerecorded and whether the microphone is used in the place of performance, outside the place of performance, or in a studio.

Claims for reimbursement of covered expenses must be submitted by the performer to the producer in an itemized statement, including all receipts, within one week after termination of employment, or the producer is relieved of responsibility for payment. Otherwise, the producer must reimburse the performer within two weeks of receipt of the itemized account. For each day thereafter that expenses remain unpaid, a penalty of $2.50 per day is incurred by the producer, until payment is received.

Work Rules

There must be a 12-hour rest period between the end of work or travel on one day and the beginning of work or travel call on the following day. Out of town jobs require a minimum 11-hour *turnaround*, or rest period, if actors are housed where the event is taking place. On days before the first performance in each city, a 10-hour rest period is allowed. There is some flexibility with this rule, which allows producers to *invade* the rest period as long as the actor is compensated.

In addition to rest periods, there must be at least one day off after not more than eight days work, and at least two days off in each 14 days of employment.

Rehearsals begin with the date the performer is first called, with no member of the cast rehearsing more than seven out of ten consecutive hours per day, or ten out of 12 consecutive hours on any three days preceding the day of the first performance. No rehearsal hours in any one week are allowed to exceed 56 hours.

There must be a 5-minute break after each 55 minutes of rehearsal, or a 10-minute break after 80 minutes. After no more than five hours there must be an hour to hour-and-a-half break.

The producer is required to provide and arrange for all transportation and accommodations on out-of-town jobs. While producers are no longer required to provide first-class air travel, the performer must travel at least the equivalent of a full coach fare. On railroad travel, the producer must provide first-class travel, or pay performer the difference. All travel associated with the job is subject to rules concerning rest periods, days off, overtime, and turnaround time.

Getting Work

Finding work in business theatre is a little different from any other area. Some commercial agents handle live industrials; legit agents who also do musical comedy may handle them; and some of the larger agencies have agents who concentrate on this one area. Learn who they are, and make sure they have your picture and resume. It helps greatly if they know you and your work. Invite agents and assistants to see you in musicals around town. Equity Library Theatre in New York City is a great showcase for this type of work. Countless people have been spotted in their talent-driven productions. Dinner theatres in and around the major cities can also be very attractive to industry people. Equity requires open calls on their jobs, so you can show up to auditions on your own. Consult the union office and the local trades to find out when live industrial auditions are being held. You'll find that the business is especially small in this area. Once you have worked successfully on one or two projects of this sort, you'll be called in to audition for others.

Interview with Carol Nadell
Industrial Casting Director
Selective Casting by Carol Nadell
435 W. 57th Street
New York, NY 10019

BEARDSLEY: Why have you chosen to cast industrials?

NADELL: As a casting director, I have more control. You are not dealing with the politics of the agency. You are not dealing with a kind of justification of position. The bucks are not the same. The importance is not the same. The people who tend to do industrials, and I am talking from the creative end, have chosen to have more of a normal life and want to have a steady income, the egos are in a different place. The same pressure is not exerted as it is in the commercial, where everything is so important and there is so much money at stake that everybody is afraid to really make a decision. They don't want to be blamed for a mistake. Their job is on the line. It's a little different in the industrial end of it. I find that you are dealing with people who want to live a normal life and have more control over their work. As a casting director who sees an opportunity for an actor to work and make some money, I have more control over my sessions. I can truly cast, which means I can *select*.

BEARDSLEY: Where do you find talent for industrials?

NADELL: All over. I find that agents are my greatest allies. You also get tremendous support from the actors. In New York if you have the commitment to act, and by act I mean act on the stage, what you get paid even in a

Broadway situation is not a lot of money. Industrials really supplement that income and allow you to have a normal family life, which I think is what most actors would like. I see industrials as supporting theatre.

BEARDSLEY: Do many actors make a career solely out of industrials?

NADELL: There are some who make most of their income from industrials, but I don't like to think in terms of people being "industrial" actors. The area where people make a lot of money in industrials is in voiceovers and in on-camera narration. The Broadway actor with film and television experience, not industrial experience, is really the kind of person I am most interested in finding. They make the material very interesting.

BEARDSLEY: You have been known to say that industrials can be "pension and welfare" for actors. What do you mean by that?

NADELL: It means they can do an industrial at scale and contribute towards pension and unemployment insurance and health insurance, which these days is really at a premium.

BEARDSLEY: Do you find that because of the recession, the dwindling financial state that we are in, there are . . .

NADELL: More actors willing to do industrials? Yes.

BEARDSLEY: On the other end, are there fewer industrials and less money?

NADELL: What I am finding is personally, knock on all kinds of wood, I am busy now—really, really busy. I find that my client base has expanded. I also find that the uses for and kinds of industrials being done are changing. I am working as we speak on an industrial for a communications company that is using 23 principals and 20 extras. It is a very interesting little film. I find that the uses have expanded. There is now a category in the Emmys for industrials. There is also in the contract, meaning the SAG/AFTRA contract, a place for extended use—foreign and cable use and even network use. I did something called "The Sales Call of the Future," which appeared on television and was nominated for two Emmys.

BEARDSLEY: Do any jobs pay overscale?

NADELL: Occasionally, they will say there is extra money. When you deal with celebrities, there is always money. Budgets may tend to be more restrictive now in terms of the recession, but you will say, "Look, if you want this person it is going to cost double scale," and they are occasionally willing to go for it.

BEARDSLEY: Are they able to cast whom they want at scale?

NADELL: Yes. But, reality? As far as I feel, a scale actor is as good as a double scale actor. I only understand double scale when they do voiceovers and they could make more money in a day. As far as talent goes, I don't see any difference. Every actor I bring in is worth overscale. I don't say, "Oh, he's just a scale actor or he's a double scale actor." Actors who [in the past] would only work for more money nowadays find themselves with their tail between their legs. They cannot afford that reputation. In terms of negotiation, if there is money in the budget, I will tell the agent and the actor.

BEARDSLEY: What mistakes do actors make on the casting call?

NADELL: Not bringing the picture and resume and blaming it on the agent. It drives me absolutely up a tree because that immediately tells me more than I want to know.

BEARDSLEY: Is there anything specific about acting for industrials that you can point out?

NADELL: With an industrial you must be able to act. For the most part you are taking a script that may not have a natural rhythm and making it believable. The challenge about an industrial is that the actor can bring something to the role. They do the same with a commercial, but the most important star in a commercial is the product. In an industrial, the actor can mold who the character is. Most of time the rhythm of each individual character is not that clear. It could be a woman, it could be a man. It is up to the actor to bring something to that, and therein lies the challenge.

BEARDSLEY: How do women and minorities fare in this industry?

NADELL: Very well, because industrials must reflect in many cases the population. What I choose to do, which is interesting in terms of industrial casting, as opposed to commercial casting where they are looking for something specific, is if I am given a role of a woman, I may bring in a minority, not because she is a minority, but because she happens to be a good actress. What I say is, "Don't make one specific role an ethnic role; let me bring in the best actors for all roles, and you will have your ethnic coalition." The actors know they are there for their talent not because a particular role is "black." I had an experience recently where they were looking for two men, a boss and a subordinate. I brought in two women and one of them was cast. In another role I happened to bring in an Asian woman who was fabulous. She got it, even though they did not ask for an Asian. The client was ecstatic, because it did not occur to them, and it added another dimension to the piece.

BEARDSLEY: Are there any other ways in which casting industrials is different from casting commercials?

NADELL: I don't think I would have the same freedom to bring in the types that I wanted in commercials. I focused on a market that is sort of a stepchild. It is not going to make anybody a star. In my experience casting commercials, games were played that had nothing to do with the purpose of the message. I like to do a job and I like to see it completed. I don't like to see it hanging. I think a lot of the casting people are subject to arbitrary decision-making: "All right, we have seen 600 of the best people, now let's see 600 more." It's frustrating and it doesn't make any sense. Nobody is competing in industrials. There are people who do them, but nobody wants to focus on it because, "It doesn't make me a star."

BEARDSLEY: What is your advice for the beginner?

NADELL: Study and study, and not just in commercial schools. Although I believe, if you want to do commercials, you must take a commercial course.

Graduate college, study acting for real, meaning theatre. Be willing to work—New York and Los Angeles are not the only places to work. If you really want to work as an actor, you will.

BEARDSLEY: Do you see any new areas opening up for actors?

NADELL: I have recently developed a division called *Internationally Speaking* because I see so much of business going internationally. It has taken me a year to find actors from 15 different countries. I constantly need to find new people. They need to be actors and narrators. They need to have a charisma in their voices. Most of the foreign work now is voiceover, but I see it also going into other directions.

10 ⬚⬚⬚
⬚⬚⬚
⬚⬚⬚

Commercial Print

Commercial print is an area open to actors and models of all types. Success in this area depends solely on your look, your ability to photograph well, and your level of determination. While at one time the commercial print business was open only to extremely handsome people or overt character types, today, as with commercials, people of all types and persuasions are finding work in this area.

Commercial print, unlike its attractive sister high-fashion print, utilizes the everyday type. Although the types used tend to be a tad more attractive than in real life, they range from attractive-real to very characterish. The industry as a whole is moving away from the yuppie image of the 1980s to a more interesting and real feeling.

As the field becomes available to a wider range of types, and as the recession causes everyone to expand their income-producing interests, more actors and agencies are including print in their commercial repertoire. While a few agencies do handle strictly commercial print, many full-service and on-camera commercial agents are including print among their services as well. Rightly or wrongly, many actors feel that a separate print agent would send them out indiscriminately, and not necessarily have their best interest in mind. The fact that their regular agent already knows them and may be working in this arena gives actors a greater flexibility in pursuing commercial print as an additional avenue of income. The agent will be able to keep the actor's exposure at a minimum while trying to generate as much money as possible.

Who Does the Hiring

While some casting directors cast print jobs, often the advertising producers cast directly or have the photographer do the screening. Sometimes the print usage comes as an adjunct to the on-camera spot, either lifted from the existing footage, or requiring a photo session, either on the set or scheduled separately.

Breaking In

Casting for commercial print is often done through photographs. While good composites are useful, they are not mandatory. A bad one will do more harm than good. A good commercial headshot will generally suffice. The danger with composites is that, unless they are lifted from actual print work, they tend to look phoney and contrived. Every young mom seems to have the same clichéd shots: one in tennis whites with racket in hand, posed with glasses in front of a typewriter, the hair-in-bun shot talking on the telephone at her executive desk, smiling maternally down at a blanket-wrapped doll posing as an infant, and so on. Like anything else, it is better to compile a composite after you have some professional jobs to your credit, instead of spending hundreds on having one made that may not make it for you.

If you don't have an agent working for you, and you wish to get some commercial print work on your own, lists of commercial print agencies and photographers who do commercial print can be obtained through the trades and other sources such as *Madison Avenue Handbook*. Send your picture to the commercial print photographers as well as to the producers at advertising agencies, letting them know you are available. Since print is unregulated by any governing body, be extremely careful in dealing with people you don't know or whose reputation you are unsure of. For this reason it is better to go through an agent. Agencies blacklist disreputable photographers. If you find work opportunities you can always refer them to an agent, or call up a reputable agency and ask if they have heard of them. If you get a bad feeling from a photographer or anyone else, follow your impulse and leave. There are many unsavory characters in this business who still do quite a bit of work, or at least pretend to. There is the well publicized case of New York photographer who was shot to death by the boyfriend of a model he had assaulted. This individual was blacklisted by many agencies yet still managed to get models into his studio. Until you know your way around, and even afterwards, try not to go to any new place alone. If you book a job on your own without the mantle of a reputable advertising agency, ask to be paid in cash on the set. Never, under any circumstances do anything that doesn't feel right to you.

Compensation

Although the print business is not regulated, the rates of pay are somewhat standardized. Most agencies have booking policies which they ask employers to follow. As a rule of thumb, the going rate for adults is $200–$250 an hour, with a 3- or 4-hour minimum. This means that if you are hired at $250 per hour with a 3-hour minimum, you are guaranteed to make at least $750. Children's rates are much lower, starting at $75 per hour, with a 1- or 2-hour minimum.

Sometimes bonuses are offered or negotiated by the agent for usage fees. A given print ad may run in one or more of the following ways:

package—appearing on the package of the product
point-of-purchase (POP)—appearing at the site where the product is sold
magazines and newspapers—appearing in one or more given magazines, or appearing in the newspapers of a specified region
billboards and transit—also known as outdoor, this means the ad will appear in billboards such as those in Times Square or on Hollywood Boulevard, on highways, and/or on the sides of buses, inside trains, subways, and buses, at stations, and transit stops. This is the area of greatest exposure, and least desirable to actors seeking legit careers.
direct mail—going directly into a specified amount of homes. This is extremely difficult to regulate.

An actor should not sign away the usage rights to an advertisement without finding out first how and where the ad will be running, and, very importantly, the duration of time including point at which the duration begins. For instance, some agencies like to begin the time span at the point of first insertion (when the ad first appears), while agents try to get it to commence on the shoot day. In any rate, most agents will not accept the job for their clients without limiting the amount of time and the extent to which the ad will be run, and receiving some kind of additional compensation. When this money is not available, the talent must decide whether the money received as the session fee will be enough compensation for the amount of exposure. At the very least, the way the ad is running and the extent of time should be limited, not left open-ended, and detailed in a contract signed by the ad agency representitive, the photographer and the talent.

The following is a rundown of a typical commercial print agency's booking policies:

Day Rate Bookings: Eight consecutive hours between 9:00 AM and 6:00 PM

Overtime Rates: Apply to before 9:00 AM and after 6:00 PM, and on Saturdays, Sundays, and Holidays

Travel Time: Charged at full fee

Preparation: Make-up, hairdressing, rehearsal, etc., full fee

Fittings: Full fee, calculated at half-hour increments

Weather Permit: Type of weather must be specified upon booking. Cancellations due to weather: one-half fee

Cancellations: Prior to 48 hours: one-half fee; otherwise: full fee. Trips and weekend bookings must be cancelled one week before booking; otherwise full fee will be charged up to a maximum of three working days

Tentative Bookings: If client does not release or confirm tentative bookings 48 hours prior to job, the agency reserves the right to cancel

Usage: Billboards, POP, packaging, out-of-home, endorsement, use of name, exclusivity, or other special use must be negotiated prior to the job, with the agency

Agency Fee: 20% on all bookings and usage fees, 2% late charge per month on invoices not paid within 30 days

---

Interview with Glenn Jussen
Headshot Photographer
First Impressions Photography, Jussen Studio
6 West 37th St., Third floor
New York, NY 10018

BEARDSLEY: How does an actor go about choosing a good photographer, what should he look for?

JUSSEN: When actors are in front of a still photographer, all of a sudden they have to relate one on one, and that is not always the most comfortable thing for them. So, the actor must feel comfortable with the person on the other side of the camera. A lot of times the actor will look at a photographer's book and think that the photos are great, but that the guy is a real jerk. If they feel that way, the photographer is probably not going to get that good a picture of that actor. But on the other hand, if they do feel comfortable they will probably let down their guard a little bit, relax, and be themselves. Then the chemistry works, and you get the good photos.

BEARDSLEY: How do you elicit a good photo from someone who is stiff and uncomfortable?

JUSSEN: By treating them as individuals and trying to feel them out as people. Some photographers do improvs, some play music, some use mirrors, some have relaxation exercises. If there were one set thing that worked for everybody, it would be really easy. But it isn't. In the process of feeling your way through with an individual and relating to them, hopefully you can find something in common that you both can feel comfortable discussing. What I always look for is a rapport with that person—loose, comfortable, and conversational—and then the camera does nothing but record that rapport. Everybody is different. So what is going to work with each one is going to be different.

BEARDSLEY: What elements should a good head shot possess?

JUSSEN: It should be an accurate representation of the individual. All right, I'm going to walk into a darkened room, I am going to spend the next two-and-a-half hours with this person . . . okay, why? What about this person is warm or inviting or sexy, or funny looking, and makes me want to say, "Yeah, I would like to know some more about them." I need to identify some element of their personality that I find attractive that I would like to know more about. If I can get that on film, then I have done my job.

BEARDSLEY: What's can an actor do to prepare for the photography session?

JUSSEN: He should decide what he wants the end result to look like. That is the biggest thing. Know what you look like, know what you are trying to market, what field you are trying appeal to, and then go out and seek it. When picking out clothes for a session, take the clothes out of the closet. Try them on, go into the bathroom, and pose a little bit. Say, okay, I'd wear this if I were going on a straight job interview or to a theatre audition or out on a date with a guy I was trying to attract or was going to meet that guy's mother. Any one of those things could work, but know what you are selling when you put it on. Know what you are projecting to other people when you wear a certain outfit. Then make sure that you are choosing outfits that reinforce what you want people to think. When the actor comes to me and knows what he wants his pictures to look like, then he and I can work together to create that end result. That's when you get the really great shots—the shots that stop people in their tracks and cause them to say, "Whoa, that's a good picture of that person!" or "I'd like to meet that actor. Let's call him in to see how he reads."

BEARDSLEY: Do you know when it is happening, when you are getting something special on film during the session?

JUSSEN: Sure, sure. You can tell when all the elements are falling into place. If I have to stand there with a young kid who does not know anything and has brought his entire wardrobe and I end up putting outfits on him—yes, I can give him looks. I can give him a young college look, a serious dramatic look, a sexy kind of soap-opera look, but whether it is the best look for that person. . . I don't know. I need some input there. Sort of like any director can give some actor a bit of business but he expects then that the actor will make it his own to do something with so that there is a give and take going on.

BEARDSLEY: When an agent, who you know likes a particular look, sends you someone, do you try to please the subject or the agent?

JUSSEN: It is always hard to know whether I am trying to please the person I am photographing, that person's boyfriend, that person's manager, or that person's legit agent. When they are trying to play it too safe, trying to have something that will appeal to everybody, you wind up with something that appeals to nobody.

BEARDSLEY: What do you do when someone walks in who is really not a glamour type but still wants that kind of look?

JUSSEN: It is really up to them. When it gets right down to it, I'm dealing with adults. If you go through the session and you get one shot where the lighting is just such an angle and your head is tilted a certain way that you look five times better than you are ever going to look on your best day, I might not even circle that one on the contact sheet. But if you come in and say that is what I want—it's your picture. If you can show up looking like that, then go for it.

BEARDSLEY: Do you find that people who do their own makeup don't come out satisfied with the shots?

JUSSEN: If you are competent at doing your own makeup, that's fine. The only time I stress it is when somebody comes into me and says they are really not good at doing makeup but can't afford a makeup artist. My answer is usually to save up. Find some way, or wait longer to get them done, because a bad shot is going to do them an incredible disservice. Any agent or casting person who gets a picture on their desk is naturally going to assume that that is as good as that person can look. Otherwise why would they send it? Trying to skimp on it for a novice actor is a big mistake. It better be you, camera ready. Hopefully it will look better than the hot day in July when you got caught in the rain and on a subway that was not air-conditioned, and then when you got to the casting director's office the 8 x 10 was there and so were 30 people already reading copy. Then you are probably not going to look as good as your head shot, you better not! By the same token, when someone yells "Places, wanted on the set!" and you walk out of the dressing room, coifed, made-up, in costume, and hit your mark, that's what your head shot should look like— so that they know what the product they are buying is going to look like when they employ you.

BEARDSLEY: What do you think of the photographers who get so hot that everyone wants their picture taken by them and all the undiscovered photographers try to mimic their style, then two years later that photographer is a has-been?

JUSSEN: I hope that I am evolving and that everyday it changes a little bit. That's all, I mean, I am always trying to keep it new and fresh. You walk that line, of course you always want to be the hot photographer, like anyone else you want to make money at this business. At the same time, I don't want to get that look that makes you super hot for about two months and then everybody is sick of your shots. I like it if someone picks it up and goes, "Hey, Elaine this is a great picture of you, who did this." I don't want, "Elaine you got a Glenn Jussen photo." Then what happens is you get second billing in your own head shot, and you start resenting it. You want people to look at you first and then ask you who your photographer was. To do that requires looking at everybody as individuals, trying to keep it fresh. I keep my schedule varied. I set up a day where I do a man in the morning, a woman in the afternoon, a five-year-old kid late in the day, or whatever. I do this so that I'm not walking in and saying, "Oh, here's 22-year-old blonde number five for this week. Remember what we did on Wednesday? We'll do that." Then you get that sort of factory approach—that grind-them-out kind of look. If I keep up a variety of people and looks then, hopefully, I see people with a fresh eye when they walk in and we can do something a little bit different with each individual.

BEARDSLEY: Can you recommend a duplicating service?

JUSSEN: *Reproductions.* They are fairly new in the business but do excellent work. I would recommend them highly. They are not on the cheap side, but cheap side pictures look that way. These people do good quality work.

11 □□□
□□□
□□□

Other Markets

While New York, Los Angeles, and to a lesser degree, Chicago, are the undisputed major hubs of commercial activity, the amount of work available nationwide has been growing by leaps and bounds. According to Mark Locher, National Communications Director for Screen Actors Guild, there is substantial work in many different regions. The actor who wants to get a start in a smaller region before coming to one of the larger advertising centers, or the actor wishing to go to a smaller market where she might find a better quality of life but still make a living, has much greater opportunity today than ever before.

Figure 11.1 shows SAG's branch offices, with number of members and total commercial earnings for 1991 per location.

According to Locher, although no decision has been made as yet, it is possible that in the next couple of years offices will be set up in Seattle, Portland, and New Orleans.

======

Interview with Nancy Lopez
Agent/Broadcast Division
Eleanor Moore Agency
1610 West Lake Street
Minneapolis, MN 55408

BEARDSLEY: How long have you been with the Eleanor Moore Agency?

LOPEZ: Three years. Before that I worked as a talent agent in New York City for several years. This is my hometown.

BEARDSLEY: What do you handle?

LOPEZ: I handle the child and adult actors for everything from a banana in a live convention to a lead for a feature film.

BEARDSLEY: As an agent who has worked in both larger and smaller markets, what are the most important, apparent differences?

Branch Office	Members	1991 Earnings
Arizona	1000	$550,000
Atlanta	700	$1.4 million
Boston	1000	$2.3 million
Chicago	3500	$22.3 million
Dallas	1000	$2,000,000
Denver	1000	$400,000
Detroit	900	$3.2 million
Florida	5000	$8 million
Hawaii	1000	$550,000
Los Angeles	50,000	$142 million
Houston	500	$400,000
Nashville	300	$1.3 million
Nevada*	700	$31,000
New Mexico*	500	$110,000
New York	31,000	$151 million
Philadelphia	1000	$1.7 million
San Diego	800	$600,000
Washington D.C./Baltimore	1300	$1.9 million
Utah**	300	$85,000

*Served through the Denver office
**This is an organizing area

Figure 11.1

LOPEZ: In Minneapolis the actors support each other. There is a real team effort
here that you do not get in the big markets. If an actor comes here from another
market, and has any kind of attitude or edge or tries to fight his way in, it does
not work. It goes out the other door for everything... with the casting people,
with the agents, with clients, with everything.

BEARDSLEY: Do you feel there is a difference in agents' and casting directors'
attitudes?

LOPEZ: Yes. Everyone here is very supportive. We all want everyone else to do
well because it makes the whole city look better.

BEARDSLEY: Some industry professionals say that outside of New York, L.A.,
and Chicago, there really is no work to speak of.

LOPEZ: I should look on my desk right now. In this past week I have worked on
a regional White Bread spot, a national Dairy Queen commercial, regional
Yoplait Yogurt commercial out of DDB Needham Chicago, Gamblers' Hot
Line Local, Folgers out of Grey Advertising New York, a pilot called "Hard-
ware" that is scheduled to shoot in Minneapolis, and *Baboon Heart*, a feature

film starring Christian Slater shooting in Minneapolis. Also a film called *Bombay*, which just finished wrapping; a couple of local industrials; another local commercial for the MTC buses here; a cold medicine test spot; Watermania, a Florida water park; *Dennis the Menace*, a feature film casting search out of Los Angeles; all the Northwest Airlines teleproductions; Armed Forces spots that run on the cable channel overseas, which include literally hundreds of spots each year; Humana Health Care, a spot produced in Minneapolis that is running in Phoenix; Image Design, a Wisconsin furniture company, Best Buy Companies, which cover the whole Midwest region, an American Cancer Society industrial, an industrial for the Radisson Hotel Systems, a feature film called *Twenty Bucks* shooting in Minneapolis in June. I don't know who is starring in that, but I know they were trying to get Christopher Lloyd. Right now I have Chloe Woodward, a little girl, testing in Nick Nolte's living room in Malibu. I had Linda Philips Palo, an L.A. Casting Director, in town last week searching for kids for *The Secret Garden* feature film that they're doing from the Broadway show. Meg Simon was in Minneapolis last week to cast for the leads in an HBO pilot. They are coming back next week to do the supporting roles.

BEARDSLEY: There are agents in New York who don't do that much in a week.

LOPEZ: I beg to differ when they say we are not doing anything in Minneapolis.

BEARDSLEY: You seem to have a lot of success with kids.

LOPEZ: Minneapolis kids are in demand. They are perceived as healthy and well adjusted. Their family life is good. They also aren't "show biz kids." Charlie Korsmo was one of our kids—he starred in *Dick Tracy*, *Men Don't Leave* with Jessica Lange, *Hook* and *What About Bob*. We also have Benjamin Salisbury starring in *The Wanderers* with Kurt Russell and Martin Short, and Anna Klemps in a lead role in *Blue Sky* with Jessica Lange.

BEARDSLEY: How big a business is the commercial industry in Minneapolis? Can actors make a living in it?

LOPEZ: Yes. We have approximately 45 exclusive actors; and, of those, more than half support themselves, their families, their homes with earnings from their commercial work. Print is very big out here. We also have the Guthrie Theatre and the Chanhassen, which is the largest dinner theatre in the country. It has four stages and supports a lot of talent, a lot of actors. We also have the Old Log Theatre, the oldest privately owned for-profit theatre.

BEARDSLEY: How about breaking in: Is it different from the bigger cities, or are the rules the same?

LOPEZ: It's not like pulling teeth. You do not have to perform miracles to get an audition, but you do need credits and experience.

BEARDSLEY: What about for commercials?

LOPEZ: For commercials it's the same as anywhere else. It's the look and it's the credits and it's the professionalism. We get 25 to 30 photos a day from people who want to be represented. The odds of booking a commercial are greater here, though. Actors may get one out of every ten or 20 auditions, verses of one out of every 100 in New York.

BEARDSLEY: Are any of those photos from people who are thinking of moving to Minneapolis?

LOPEZ: Most are from people who live here. The people who are thinking of moving from larger markets with credits have no trouble getting an appointment. They need to be real relaxed about it, because we are always looking for good talent. There is no way for anybody local to get the credits that someone can get in New York or L.A.

BEARDSLEY: So just pick up the phone and call the agent?

LOPEZ: Send your picture before you get here so that we can expect you, and when you get here walk in with it.

BEARDSLEY: And what about somebody starting out in Minneapolis and moving to one of the big two or three, is it any easier for them?

LOPEZ: I think they can get some skills here so that they can be prepared once they get there, but it is not going to get them in any doors quicker unless they have been at one of our major theatres. It might for commercials if they have really been successful. Getting into the unions is much easier here.

BEARDSLEY: Are agents or casting directors less strict about not taking calls or walk-ins?

LOPEZ: You don't want to interrupt anybody, so it's best to send pictures and resumes and do it the polite way. But the point is, once you do meet someone you don't have to push for them to like you or to watch your tape because if you do it turns them off. You need to be really open and receptive about it, and if you cannot get an appointment you go about it the same way that you would for anyone else. Get on stage, be seen, the whole thing.

12

Actor to Actor

The following interviews are stream-of-consciousness thoughts from actors in the field. The subheadings of "New York," "Los Angeles," and "In-Between," represent where they are physically as well as where they may be in terms of their careers. While some have attained what their peers may call success, and others are still groping at the rungs, all have been successful at developing themselves as individuals. The process of being a "working" actor never stops. The struggle is always there, no matter where you are in your career.

New York

Kathleen McNenny

When I got out of Julliard I was a "legit actor" who was supposed to go out and support myself. But the reality is that only a very small percentage of our profession can support themselves entirely on legit work. So the question becomes, "How can I support myself between jobs, have the flexibility to go to auditions, and still maintain some sort of creative integrity without becoming trapped into a nine-to-five job?" I decided a commercial career was the answer.

I needed something that kept me from becoming a desperate actor, one who *had* to have the job. When you go into a legit audition desperate instead of going in relaxed and comfortable, they can read it all over you. The hitch was that I had just completed four years of an intensive classical training program that included almost no camera experience. My first commercial audition was for a Chicken of the Sea spot where you eventually were going to wear a mermaid tail. So we were asked to wear bathing suits for the audition. Well I showed up—and I'm not kidding you—all the Ford models were there. I go bopping in in my Speedo, and the casting director says, "Slate." I just looked at her with this blank expression because I had no idea what a slate was. She stuck her head out from behind the camera and said "Say your name." Then she said, "Three-quarters." I still didn't know what she was talking about. Then she said, "Hands," and I looked down only to discover the box of blueberries I had just eaten were caked all around my cuticles. So I went out chuckling, "Maybe I better learn something about what I'm doing here."

I signed up for a commercial class. My instructor, Ruth Lehner, was great. I got experience in front of the camera, the chance to see what I looked like, what to wear, and so on. I booked my first job while I was still in class.

People shouldn't give up on commercials too quickly. Success is a matter of volume. They need to really stick in there. The more auditions you do, not only the better you get, but since it has a lot to do with luck, the odds start changing in your favor. I had literally been out four and five times a day every week for months and months and months. I became the first-refusal queen for a while. My first year I booked one job, the second year I booked two or three jobs, and then the next year, all of a sudden, *boom*, one or two a month. I think you've earned your money by the time you have the year I finally had. I don't know if all of a sudden my look was in, or that the casting people had seen my face enough to throw in an extra good word, or what. But I started booking, thank God!

The turning point happened after my second year. As my commercial career became more successful it provided me the financial security to change my environment. I was able to move out of Hell's Kitchen and into a nice apartment, without constantly worrying where my next dollar was coming from. So I moved, and my whole life changed. That happened because, I believe, of a new apartment. It's someplace safe and bright where I feel happy. My other place was scary; it was oppressive and dark and the neighborhood was awful, but that's where I would have had to stay had I not made commercials. Because of commercials my environment was better, my legit career became better, my whole life was better.

Commercials gave me the financial freedom to be able to wait for the good jobs, I didn't have to say "yes" to every little job that happened to come by. I could say, "No, I don't have to do this reading for $50 a week because I desperately need $50." That's the other advantage: Commercials allow you to be a little more flexible.

Kathleen McNenny is an accomplished stage actress, who has also worked and starred in feature films, television, and many commercials.

Patrick Quinn

There are no why's and wherefore's in the commercial business. Commercials don't seem to make sense, whereas theatre makes sense to me. While in theatre you get a job because of your talent, in commercials I sometimes wonder if that's true. I once had a client come up on a shoot and say to me, "You know what we really loved about your audition tape?" My ego was ready to blossom. "That blue sweater you wore. It was so wonderful, it made your eyes just punch out on screen." I thought, "It had nothing to do, perhaps, with the way I read the copy?"

When you are on the set you are really one step below the product. You have to keep your wits about you and your self respect intact, because you are in many ways inconsequential to what is going on. That's what I found to be the most difficult thing to deal with. I had always been respected for my talent before [commercials], and it had less to do with that now for some reason. That is not to say that talent isn't important in commercials. I really think it is, but it doesn't seem to hold as much importance. I'll walk in on an audition or a callback, which now seem to be "all-backs." I remember when first refusals used to mean three of us; now they are so abused that it's 25 of us plus all the new people they want to bring in on that session. I may see ten guys I know there. I know a lot of these guys now because I've auditioned with them or because I've seen them on-camera (that's the thing about commercials, you're constantly seeing what you've auditioned for and may not have gotten.), and I know how talented they are and that any one of them could do that commercial. It becomes a lottery; the rules have broken down.

I was told constantly by friends that I was a great commercial type. I think everybody and their mother is a commercial type. Commercials cover such a broad spectrum that there really is no such thing as a commercial type. The reason I think that I've been able to be so successful is that not only could I do "spokes" (straight spokesman with jacket and tie), but I could do the off-center spokesman as well. Most of the stuff I've ever booked has had a comedic touch, or was character stuff. My agents have always been able to send me out on a variety of things. I'm not a one note person when it comes to commercials, and that has helped me a great deal. But in the last ten years, because of the dearth of theatre and the recession, actors in my age group have had to be resourceful. They have more financial responsibility because of children and homes and other things. A lot of people who would never have done commercials (and I'm not talking just about stars, I'm talking about good, respected theatre actors) now do them because they have to pay the rent. It's like soaps—they used to be below everyone—if you did one you would be dead in the water for the rest of your career. Now they're stepping stones to other things, and so are commercials.

I think that the talent pool has grown tremendously because of people realizing that they just can't do film, or do theatre, or do TV, or do commercials. You have to do everything. Now if I could just figure out where.

A veteran of dozens of commercials, Patrick Quinn starred in the television series "Bosom Buddies," the Broadway shows Oh Coward, *and* Lend Me a Tenor, *and is vice president of Actor's Equity Association.*

Ralph Buckley

I was a baseball player. Even the best ballplayers only succeed three out of ten times. The ballplayers who are considered to be the best (.300 hitters) fail seven out of ten times. If you dwell on your failures, it just eats into your spirit. I try not to dwell on what I don't get, but to think about what I do get and let it feed me. I have enthusiasm for whatever I do: a radio commercial or a network spot, off-Broadway or a play in New Jersey. I try not to put value judgments on anything, but see it all as work.

Ralph Buckley is a self-described actor-of-all-trades who does a little bit of everything.

Yvette Edelhart

I am basically a very sane person who has not ever hit rock bottom. I am fortunate in many ways. At my age I am living here without a lot of responsibilities. If I make a lot of money, or don't make a lot of money, I manage, because I don't have huge expenses. Careerwise, I enjoy studying and learning, and that's been very helpful to me. I am now in charge of a group called Professional Actors Workshop. It's an actors' cooperative, and we bring in people to hear our monologues and to possibly use us at some later point.

I think I have accepted certain realities about myself in the business. I will not be a big Broadway star. I will do very well commercially, as I have done. I have learned to accept some things that are a little harder to accept than others are. I don't even let not getting a commercial bother me anymore. They either want me or they don't.

I started this late in life, which I think is very helpful. I didn't begin any phase of this until I was in my late 40s. It started as a hobby and grew into a career. I still find it exciting and fun, and I enjoy it. I enjoy living in Manhattan Plaza, and that's very nice because it's very safe. I have acquired two cats. The temptation for me is L.A. Should I try it? Am I losing out?, and I wonder, then I say "Look, what do you need from L.A., a huge career? You don't need that." I'm a big believer in fate. If it's going to happen, it's going to happen. You have to be prepared when it does, but don't push. For my 64th birthday, which just passed, I had my cards read. They said something big was going to happen at the end of this year.

I got my start in Chicago. I got my union card doing a dinner theatre production of *The Mousetrap*. I went out on my first commercial call through my agent in Chicago, Harrise Davidson, and I booked Illinois Bell. I had never done it before, I had never been in front of a camera. It just came naturally, and I was simply a new and interesting face.

One of the best pieces of advice I got was from my mentor in Chicago, Bud

Beyer, who said the minute you get there [New York], put your name on the list for Manhattan Plaza. I also learned the hard way: Don't spend your money until you've got it.

In New York for over nine years, Yvette Edelhart has made her SAG pension working as an actress in commercials, TV and film.

Kate Weiman

Since 1973 I have supported myself as an actress, so I've never had to worry about survival. For mental survival I do other things: I like to paint and do headshot photography. I have been lucky in that I have approached my career by always working on it; I never have grown tired of my career. I'm always doing something new, getting new pictures, taking a new class, staying fresh with theatre.

My career is a reflection of the rest of my life—photography, tennis—in fact I just did 24 radio spots for Ryder Truck Rental through my tennis partner who works for Ogilvy and Mather. I don't rest on my laurels, I don't take what I do for granted, I am always out there. Like any salesman, I have to be out there keeping active and interesting.

When I turned 40 I felt I had to go out and shake things up. I did standup comedy, which loosened everything up. All my activities help with all the other activities, not just acting. I do my life fully. Right now I am taking a computer course, not word processing, but something called "Photo Shop," a program that can take any photo and manipulate it. This is something that excites me, not something to fall back on just in case I'm scared. It's a skill that I want because it will enhance my photography, and I enjoy doing it.

The biggest lesson I learned was to break the rules as much as possible. There is a tendency by beginners to want to fit into what they think the agent or casting director wants to see. I wish someone had said to me, "Go as yourself and be all that you are." Know that your uniqueness is special. As soon as we second guess what others want to see, we are in trouble.

Kate Weiman is an actress, comedienne, singer, photographer and tennis player who has appeared in feature films, theatre, television, and numerous commercials.

David Fonteno

I figured out somewhere along the way that what I do now for a living would be the only thing to sustain my interest and passion for the rest of my life. This business is the type of business where resetting goals is a constant: When you finish one project you pursue the next. When you reach one level of success there is another to be achieved.

When I was young I just wanted to work and to be a star. Now that I'm working I'm not a star, but it's not important to me now. What's important is doing the best possible work that I can do. As I approach middle age I'm thinking about income, too.

David Fonteno is a corporate spokesperson for United Way, in addition to having dozens of commercials and other industrials to his credit. He has had contract roles on "Search For Tomorrow" and "Santa Barbara" and works regularly in theatre.

Betsy Friday

I chose in my career to keep my options open in order to prevent hitting a low period. If a commercial wasn't hitting I would try for a play, musical, or industrial. I diversified, and the different areas complimented each other. No one job becomes too important. We are in a service industry. As an actor, our talent services the vision of the creative team, the client, and the producers. That is our creative contribution. In auditioning, I firmly believe you have a 50/50 chance. Either *you* get it or *you* don't. Information regarding the project, training, experience, and even nepotism contribute. But it is that moment you present yourself before the creative team that creates the possibility. They want you to be what they are looking for when you walk through the door. You must do your homework. Leave as little to chance as possible. Have no excuses. Prepare the material asked of you as completely as possible. We are our investments. If you are making this profession your career, think long term. Every audition and job develops relationships. You can learn from everything if you let yourself, which is really the fun of it and why we do it.

Betsy Friday is an actress, singer, dancer, currently in the Broadway production of The Secret Garden. *She recently served as the assistant choreographer for the national tour of* The Secret Garden.

Jeff Phillips

At 18 I found getting into this business difficult. I auditioned for 30 or 40 commercials without getting a callback. Forty might as well be a hundred or a thousand. After a while it got tougher and tougher to keep going out for them. It seemed like I had to get past some sort of threshold, past all the times of not even getting a callback. Then finally I booked an Army spot, a national. I remember, because I was back home in New Jersey, cutting my folks' lawn. I remember thinking "I'll never have to mow lawns again." That spot still has yet to run.

Booking that spot gave me confidence. What it told me was that anybody can book these things. I was able to see that there were a lot of variables involved that didn't really have anything to do with me. Knowing that made it easier, and I was able to bring back the fun, as opposed to dreading yet another humiliating audition.

About at the same time I started doing some extra work on some soaps and getting myself into plays. I did a recurring on "As the World Turns," then I booked another spot, and another. That year I ended up booking six nationals, and 0 for 40 turned into 1 for 6.

Jeff Phillips plays Hart Jessup on "Guiding Light," and has been nominated for a daytime Emmy.

Terry Gatens

My first commercial was Miller Beer with Danny Baldwin and Rob Morrow from "Northern Exposure," and it was just sitting on a couch drinking beer watching a football game. It wasn't too hard. Any one of 2,000 guys could've done it, but the fact is, I had the look they wanted. New faces are great. A lot of people will agree that the longer they are in the commercial business, the fewer jobs they are going to get, because their face has been seen and commercials are about slice-of-life. With the exception of me doing this Dr. Pepper thing, or the Orville Redenbacher kid, or the Maytag guy, commercials really want the consumer to believe that these are just average people. It pays to be a new face, and all the more reason to pursue it at the beginning of your career. Commercials open up so many doors for you; they give you your tapes, money, and exposure—which is the most important thing, to be on a set, to be learning. They give you contact with casting directors who are going on to do other projects. I don't think anyone should ever put them down.

What helped me a lot was improv, it frees you up. I studied with the Groundlings. There's a technical aspect to commercials, meaning you can't turn your back on the camera because no one wants to see the back of your head, but there is also a freeness that people casting commercials want to see—just high energy and improv, being quick on your feet, throwing yourself into situations and making them interesting and funny. So many times you go into a commercial audition, and they say, "Okay, you're on a beach, and you're walking with your wife." I ask for a couple of seconds. It's your time. Even if it's only two minutes, it's your two minutes and you make it your best two minutes. Ask the casting director, "Could I just have a second here?" Confer with your partner. Say, "Look, we're on a beach. This is our first weekend away, so we're kind of nervous about it, but we realize that we love each other a lot, and we're on a *certain* beach, at a *certain* time of day. A lot of casting directors will object, saying "There's a lot of people waiting," but you've got to say, "I want just a second here. It's my time."

This business is not about rejection. There are so many other factors involved in making it other than you the person, especially in commercials. If you keep you, the person, centered and strong, then everything else just flicks off you. You just deflect it. No one should tell you to fix your nose, or to get bigger boobs, or dye your hair, or do anything like that. You should never have to do that. If you try to become somebody else, you take away from your uniqueness. This business will manipulate you and try to make you feel like less of a person, and it is not about that at all. A lot of actors are damaged, and I get damaged too—you have your low points. But you have to realize that every low point is as low as it is going to get, and then you start going back up again. When I first got to the city and booked this Miller Beer commercial, I thought "Great, I got this by the balls." I moved into Spanish Harlem and I sat there for nine months, tearing my hair out, working three different jobs, asking, "What am I doing? I have a college education, I graduated with honors in business." But then all of a sudden I got with my agent, and it became numbers. They had me on five or six auditions a week. Before I knew it I had six commercials on the air. Then I had the financial flexibility to sit back, relax, put things in perspective, and pursue a legit career. I never could have done it if I didn't get the commercials.

Everybody at the beginning of their career has done what they had to do to make money. You have a product, and you have to do whatever makes this product marketable. Anything that diminishes the product diminishes the process. And like any other business, you should be showing a profit at the end of three years. You should be getting something back, and if you're not then you have to think about what you can do to make your product more appealing to the consumer.

Known to some as "the Dr. Pepper guy," Terry Gatens works in theatre, film, television and commercials.

Los Angeles

Katie Mulligan

When I reach the point where I'm not booking jobs or getting the callbacks, I take a good look in the mirror. I give my life a once-over to see if I'm physically up to par, including what I'm eating or drinking. Am I also mentally centered? I take a look at what's selling on television, in my age groups, parts that look like I could play them. Recently I cut my hair and permed it. I look at what they have and what I am, and make some decisions. Then I'll have a talk with my agent, casting directors who know me, or a commercial instructor, to see if my readings are off. Recently I asked to see a playback of my reading and saw that I was off-course. Check to see, and decide whether you want to be on-course or not. That may involve more classes or revamping a physical aspect; it could involve going to the doctor or chiropractor, anything therapy-wise to get you back on track. It could be a small thing: Once I'd tried everything else so I changed the color of my lipstick and started booking. You have to see what may be affecting you. Changing my lipstick may have changed the way I feel about myself.

Katie Mulligan is an actress who works in theatre, film, television, and commercials.

Mary Gordon Murray

My agents just kind of drafted me commercially. I never avoided them; I just had always worked doing theatre, and I had a soap opera job, and money had not been a big problem for a long while. They were wonderful; it was such serendipity. They always sent me out a lot on-camera, and I would work, but mostly what happened that was wonderful was that I clicked with the voiceover work. I never stayed in New York long enough to really cultivate that as much as I could, because if I really needed to make some money doing voiceovers I'd have to just stay in one place. I'm always running off to do a play somewhere, or moving to the coast or whatever.

Even while doing the work that I'm doing here in L.A., something I tested, a Jello spot, went from a test to a major national network and it's just pouring in money. I did it three years ago when I was in New York for maybe four weeks.

For me as a theatre person, I liked showing up. As silly as it may sound, commercials are something that you go up for a lot more than plays or films, and I didn't realize how much that anchored me, it kept me grounded and focused on something. One of the things that I really had to go after here in L.A. was commercial work, and now I'm just starting to have that happen for me here. I didn't realize how much I counted on that, even for just the bread and butter of it. I'm lucky, I've always managed to have enough work to live. But commercials are like a backbone that gives me enough stability to continue to do what I want to do, which is not take jobs theatrically that I don't want to do, and for the last ten years of my life I really haven't had to do that. That's a combination of good planning and commercials.

I've never been someone lucky enough, or just by design or desire, to be very active in commercials. Even if I wasn't booking all the time, the on-camera work amounted to usually just a few a year. I came really to depend on voiceovers. I loved it. I had a much better batting average with that. But that was the backbone of my work week. It was a much more important part than I realized until I came out here and didn't have it the same way. I thought I'd miss the musicals and the readings, but that's the one area more than anything that I took for granted—just those three appointments a day.

They handle it differently here. I think it's more separated. There's a funnier notion about commercials, I think, in L.A. There are people who just really don't do them. In New York, nobody cares, you could be Dana Ivey or whoever, and it would have nothing to do with your credibility as an actress. I've actually heard people who are hesitant about doing too much on-camera, because there's a snobbery about it. They're probably the same people who would tell you not to do a soap. I think it's mostly hogwash. If you are dying to be respected, then make sure you are doing theatre at the same time. Don't *not* work, just do *more* work. The only real stymie of any of this is if you let it stop you from doing other kinds of acting, if that's what you really want to do. I've always heard that excuse about soap operas: you'll get into a rut. If you sit back and get your house in the country and don't worry about doing any theatre, then you'll get in a rut. The snobbery doesn't even exist out here as much as it did once. People are just much more practical, and the business has started bleeding, it's not really as segregated and as specialized. It's work. That's healthy.

Mary Gordon Murray was nominated for a Tony for Little Me, *and works steadily in theatre on Broadway and across the country. She does work in films, has a recurring role on "L.A. Law" and for five and a half years was known to viewers as "Becky Lee" on "One Life To Live."*

Barry Kivel

Out here in California you basically have to start all over again both theatrically and commercially. It's not just California. I noticed a change even before I saw the volume taper off. For some time I would do between 12 and 16 commercials a year, and I knew if I didn't do those one of a select group of actors in that same category would show up on the TV. Now I don't see the same actors that I used to or the same types of commercials. The market has changed.

I came out here to do television and film work. I had established myself in New York, which meant directors who had clout would fight for me as their choice. That was my advantage in New York. I was giving this up with my move to California.

It's a long journey. Film and television are hundredfold more difficult than commercials because the stakes are much higher. Needless to say, one would like an instant career, but it is not always smooth sailing. Who said it would be? There are ground rules as in any other business. To be an actor one must *commit* and *persevere*.

A final note: Don't judge your career by anyone else's career. Yours alone is unique.

Barry Kivel has a recurring role on "Civil Wars," is Murphy Brown's *network lawyer, and has appeared on "Star Trek The Next Generation," "Sisters," "Empty Nest" and "Full House." His film work includes* The Natural, Crocodile Dundee, Soapdish *and* Memoirs of an Invisible Man.

Leigh Curran

Strangely enough I have a much better look out in California because my hair is gray and my face is young. For me that happens to be working very well right now because there are very few actresses who have let their hair go gray. I think commercially people like the idea that you are a healthy middle-aged woman, and that is sort of what I am communicating. I wasn't sure when I came out because everyone dyes everything and surgically removes everything. I thought that I might have to do the same, but I was not willing to do that. In my heart, I just had a feeling that my hair would work the way it was and it really has. I also think I am a better type out here than I am in New York because I communicate a kind of warmth and openness. In New York they like high strung, anorexic women—or anyway, that has been my latest theory. To get myself going there I wrote myself parts in plays and got them produced. Commercially, I still play moms and businesswomen. The only problem is that there are fewer roles available to go up on, and of course there are tons and tons of people out here. So you are up against three-quarters of the industry. Your physical type means everything out here. Mine happens to be working for me, so I don't have a particular ax to grind, because I am also a good enough actor that I don't feel what I have to offer is all about my body. To me that is the way to be in Hollywood. I think it is deadly to be here if you are only pretty, because you can't develop a career; instead you develop this fear of age, which is like developing a profound fear of yourself.

The problem I run into more often than anything is that I am really a character actor and I am also good looking. People get very thrown by that because if you are a character actor you are supposed to be kind of homely looking. Often, I was thought of as too pretty to play certain kinds of parts. I think that actually that is not as big a problem for me out here. People in L.A. *like* you to look good, so they don't care if you are playing a bag lady and you look good.

Being out here has worked for me really well. Although there is not a whole lot of work for women in my age group, I am immersed in *The Project* so I don't really care. I think the real trick to being an actor in this day and age—and this also applies to actors going out commercially—is to do other things so that all your eggs aren't in one basket. You are not counting on having to make *X* number of dollars a year in order to survive. You have to find all kinds of ways to get by. I basically do three things, four things including commercials. I act, I write, I am an artistic director of The Virginia Avenue Project, and I support myself by making commercials. I really have fun making commercials. They represent this little mini challenge. I like it best when the commercials are improvised. I don't have any trouble when they call me up and say come in and talk about this or be that. That to me is the fun part.

Leigh Curran is an actress, author, and Artistic Director of The Virginia Avenue Project, an organization that puts inner city kids together with professional writers, performers, and directors to create theatre.

Randy Rudy

I fell into an agent by accident, by walking up and down Sunset Boulevard handing out pictures and resumes. I didn't know that's not really how you get in. As far as getting into this business, people need to realize that, number one, there are no rules. What works for one person will never work for another. It's just as possible for the guy pumping gas to be a major movie star as the people who are doing really well, but may never work again. Number two, no matter what anyone tells you, including Randy Rudy, God, and William Morris, it's only one person's opinion. I've had people tell me I'll be a really big star, and I've had people tell me to get out of the business.

I had the head of casting for one of the networks suggest I really needed a "hook." His idea for me to be successful in the acting business was to move to Chicago for three years and get with Second City. I thought that that was possibly one of the stupidest things I'd ever heard in my life and luckily didn't take his advice. But this was a person who could've given me a job or started me in a career. His idea for me to make it in L.A. was to leave town. It doesn't matter who you are talking to, it's just one person's opinion.

Occupy your time with something productive so that you feel like you are getting something accomplished instead of sitting by the phone and worrying why you are out of work. I have a house I've been working on for the past ten years, which has been salvation for me, good therapy, and also a good investment. I used the money I made in commercials to purchase the house, and then I put more money into it working on it. Have something to occupy every day, whether it's changing a light bulb, building a wall, or adding a bathroom. When you hold a beer can up all day and say "It's Coors Light," it's not exactly terribly fulfilling.

At this point in my career, 12, 13 years into it and 150 to 200 commercials later, 50–75% of my business is repeat business—directors, production companies, or advertising agencies I've worked for. When they spend 75, 100 or 200 thousand for the day, they would just as soon have somebody that they know is reliable, who shows up on time, says the right thing, and wears the right clothes. It's like any other business, be nice to the right person, say the right things to the right people, and at some point have a sense of humor about the whole thing and not take it too seriously. It's a combination of many emotions all at once.

Seen frequently on commercials, Randy Rudy's passion is building furniture and doing home restorations.

Taylor Nichols

The turning point for me was when I took an on-camera course from Anita Guerrera. I had taken acting classes with Terry Schreiber for years, but Anita's class was the first real tangible output—the first chance to see myself on-camera. She did not make it "fancy-dancy, just-smile-all-the-time" commercial stuff; she allowed us to *work*. She did say, "Now yes, this *is* a commercial, so be upbeat," but she also allowed us to feel, or emote, or do whatever we wanted to do. And through her class I met my future agents. That was a really important step for me. When and if young actors ask me for any kind of advice, I always say, "When you are ready, take a commercial course." That's a good start.

In comparing commercials on the East and West Coasts, it appears to me that commercials on the East Coast are more dialogue- and actor-oriented, and commercials on the West Coast are more image-oriented. When I'm in New York, I go out a lot on commercial calls and I get called back a fair amount. When I'm on the West Coast I don't go out as much, maybe on one or two a month. There's hardly ever any dialogue, and if there is, you just improvise. It's not set on a cue card, where you actually get to practice and show who you are through the lines—you have to show who you are while walking around looking at this cool car. Beer or car spots are the things that come to mind out here. I just don't think I personally can sell myself fast enough just on my look—I need a chance to have the camera look into my eyes and work a little bit.

It seems to me here in L.A. that the actors I see on commercial calls are either models or real character types, whereas in New York you see all kinds of actors showing up on commercial auditions and industrial auditions. Out here it's more split between commercial actors and models and "actor-actors."

When I first came out with the TV show, I talked with my agent, and we came up with this plan that I would not go out on commercials unless they were really good spots, not the spots where you hold the product up by your face and smile, and not hemorrhoid spots or such things as that. So when I first got to L.A. I wasn't going out very much at all.

I miss the income of commercials. They were great for me—it was exposure. I can't think of any actor who was hurt by overexposure in commercials. Maybe a few have, but with the way commercials are going, more and more now to real actors, and real acting, and honest moments, you see a lot of actors move from TV commercials to television and film.

A lot of actors look down on commercials. I think that's a mistake for a couple of reasons. One, doing commercials gives you a lot of experience and confidence in front of a camera, including learning to do stupid things and making them real. TV sitcoms are not that far away from certain commercials; lots of times commercials are better than trash sitcoms. Second, the money is great. And often you work with good directors. I've worked with Ridley Scott's company and Bob Giraldi. The Ridley Scott commercial was beautiful. The subtle little acting moments that we had in the spot were exciting. In a way it bothers me that we were promoting a product, but at the same time, who cares? I got paid to do it. And I got to work with good actors and a good director in France, on creating a nice 60-second moment. So people can look down on it if they want to, but I made a lot of money on it, plus it was fun.

Taylor Nichols had starring roles in the film Metropolitan *and the television series "Man of the People" with James Garner. He recently made the film* Over Easy, *which was featured at the Seattle Film Festival, and has been seen in many commercials.*

Jonathan Penner

My understanding of the business is that there are two stories. The first is you are fresh out of school and have all of that energy and training and drive and passion, and sort of burst upon the world and immediately get an agent and start getting work, and there are no down times until much later. This is true commercially and legit. I have a friend who just graduated from Sarah Lawrence in May, got an episode of "Law and Order" in New York, and she got my show in L.A., and then she got her own TV show, all within a year of graduating.

Or, if it doesn't happen like that, if that energy and drive when you first hit the scene doesn't pay off, and you sort of find yourself after six months still unemployed and still without an agent, then it basically takes three years to get that next break. That's the other story. You have to sort of pull yourself together and you find yourself in the trenches, waiting tables, or doing a day job. If you *really* keep at it, and you *have to* keep at it, the time span that I keep coming across is three years before things start to pay off, and that's a tough slug. But if you look at almost any other career, it's the same way. If you were to become a doctor or a lawyer—or something much more banal—either you're a superstar and *boom* you've joined a great company and you're set, or you've got to work for some yahoo and do a lot of the crappy work around the office to get experience and learn the game.

You are definitely going to get a lot of rejection if you haven't just exploded onto the scene, which is not what you are trained for. You are trained to know what to do *when* you get the job, not to *get* the job, and to deal with unemployment, and with pounding the pavement as they used to romantically call it. I never thought of it that way, but looking back it makes sense because you can't afford cabs, so you're taking the closest subway from one appointment to the next appointment. I used to literally walk to appointments because I'd rather save the dollar than spend it on the subway, and the one thing I really had to spend was the time.

After you've been rejected 100 or 50 or four times, you have to say to yourself, there's got to be another way of looking at this so that it doesn't shatter me. I decided to become an actor. I *am* an actor. How do I deal with the pain of trying to get a job? It's going to take a while for everybody to know who I am—letting the casting director know who I am so that if I don't get the job, the next time, or six times down the line they'll say, "What about that guy...?" Then there's meeting an agent, because the first time you meet an agent, yeah they'll send you out a couple of times, maybe, if they see something there, but part of the pressure becomes to deliver on those couple of auditions so that the feedback makes them say, "Hmm, I guess I have to keep my eye on this guy." That's part of why you can't take it personally. It's all a business, and while this is your only career—this is the only career that you need to be worried about—all the other people on the other side of the business have a *lot* of people that they need to be worried about. They are all looking for somebody who can make them some money, somebody who can make them as a casting director, because they discovered the next whomever.

What was very helpful to me was to remember that they *want* you to be good. You walk into an agent's office, they want you to be great—a wonderful type, or a good personality, or something—because if you are, then they've got you. They are going to make some money. They are going to look good. It makes their job a hell of a lot easier. The casting director wants you to be the guy who walks in and blows them away and is exactly right and perfect because it makes their job much much easier. They can point to the tape and say, "That's the guy," and the director will say the same thing, and *boom*, you've got the job. If you go in thinking they're the enemy, they're the bad guy, they're the guy who's going to reject you minutes after you walk in the door, you are going to come in with an attitude that is going to make them do just that.

Jonathan Penner played the role of David Strauss on CBS' "Grapevine" and has appeared in the films Religion Inc., White Palace, *and* Amityville 1992. *He is also a screenwriter.*

In-Between

Rosemarie DeBlasi Garner

I wouldn't want to go back to being an actress in my 20s because of the person I was and how intimidated I was by people in positions of power. I always tried to second guess what they wanted. Now that I'm in my 30s, I don't think that way anymore. I feel a lot more secure in myself. Of course I have terrible moments of being depressed and thinking nothing's going to happen, and I know that will come again because I'm realistic. But right now I feel a much better sense of who I am and can meet people on an equal level. As a result my work is much better.

I also now have a more realistic outlook as to where I stand in this business and what pieces I could seriously present to people. How I view myself now is probably how they view me, not who I *wanted* to be. In my 20s I wanted to be someone with long blonde hair—that's the ideal. But now I don't want to be that person. My work is a lot better because I'm not trying to be what everybody wants me to be.

It's easier to schmooze now. That's a big part of it, not that I think I have to compromise myself and be phony. I just feel that because I know more who I am, I feel much more self-confident. I can go up to someone in power and be witty and charming and not feel that I'm being someone I'm not. I don't feel as I used to feel that I have to make myself invisible because I didn't think I was important enough or worth anything—why would they look at me when they have the girl with the long blonde hair standing right there. I have other things to give. Until you come to that point yourself you can only understand it in an intellectual way. Successful actors say, "I reached a point where I didn't care anymore, but I still cared just as much," and they exist at the same point. I still care as much, but I don't care.

Rosemarie DeBlasi Garner can be seen frequently in New York theatre and on various soap operas. She keeps her talent fresh with numerous classes. She and husband Patrick Garner are planning a move to the West Coast in the near future.

Tommy Koenig

When I first became an actor commercials were on the bottom of the list because I thought of myself as an *actor*. I was into experimental theatre. Commercials became important to me when I first got to New York and was really struggling and waiting tables and trying to be an actor with an underground troupe. It became frustrating, and I said, "Maybe it's time to give it a whirl." Serious actors had a cynicism about commercials: They were considered below us. A *real* actor doesn't do commercials. But it became a matter of finance. If I want to be working as an actor I have to be able to survive, but I also have to have visibility. That's the thing that commercials give you—that instant visibility. People can actually make their career on commercials, and at the same time it can also kill a career. That's the fear I had going into it. What if you get in the kind of commercial that sort of pigeonholes you in something for the rest of your life?

I found that I was accepted in the commercial world and I learned how to love it. In a way, that's how you have to approach it. You have to love what you say and love the product. Sometimes that's *re-e-ally* acting. I said, "Wow, this is some of the biggest acting I'll ever do." I found them to be fun, and as long as it was a product that I believed in and didn't feel some embarrassment about, I figured everything would be okay. You're often given that choice. There are people who won't do beer ads because they feel it's a bad influence to do things like that.

I also like the activity of commercials. It's one of the places where an actor can work a lot. You can do a play, and rehearse for months and months and months, and then perform for maybe two weeks. Or you do a series or a pilot, and you get a one day episodic or a week on a show, or a couple of months in a movie. It's a constant, a constant way of keeping your skills honed and being out there as an actor and getting seen and hopefully developing a persona that people want to see more of. I do notice that when they fall in love with somebody, they just sort-of eat them up. I've never gotten on that sort of roll—I'm working on mini-rolls in that I work sort of steadily—but you see people who are in just about every commercial.

There are times that a commercial calls for a certain type of performer or a good interesting character or something eye-catching or interesting, and I think this is where an off-beat person such as myself can come through, especially in comedy. These are places where I've really gotten to excel—when they need somebody to play a coupon, or a credit card, or a talking telephone, I'm your man. 'I do a great tamata, I do a whole series of vegetables...' That's something I can do. I've always done the off-beat kinds of parts and there are a lot of those available in commercials. You get the same amount of money, and you don't have to be overexposed.

Tommy Koenig is an actor-comedian who works quite a bit in film and television. He performs his one-man show in theatres and cabarets and headlines comedy clubs nationwide.

Patrick Garner

I run into a lot of legit actors who think if they could just get one commercial everything would change in their life. That's not true, mostly because the commercial business changed about seven or eight years ago. It used to be that if you got a national network commercial, it ran for two years. You might make $80,000 on that one commercial. That doesn't happen any longer because they changed the way they produce. They do a lot more commercials, which means even though they spend more money, it's going to a lot of different places, not to one specific actor. If they do five commercials in a year, they're using five actors, as opposed to doing one commercial using one actor. You don't make the money you used to in commercials. I've found that out because I've only been doing it three or four years now, and at times when I've had a lot of commercials running, people who have been in it ten or 15 years say, "Boy, you must be raking it in." Then they catch themselves and say, "Oh, but we don't make that much anymore." So that's the caveat I'd have for other actors: not to look at commercials as a way to solve all their money problems.

When I first started out I had to put aside a year to really pursue it. I was engaged, and we were trying to save up some money, so I decided not to go out of town and do regional theatre, and not to do showcase work, because I wanted to concentrate on commercials and get known with the different casting directors. It does take awhile to build up a relationship with them so that they know you after having been in a few times.

I couldn't get a commercial agent for six or seven years. I would meet an agent, they would send me out once, they wouldn't hear anything, they'd drop me.

When I auditioned for the K-Mart spots—they were originally casting five 15-second commercials—I went down to the casting director's office and the assistant assigned roles according to what you looked like. I read in a room about four feet by four feet. The camera was in one corner, you and your partner in the other corner. It took all of two minutes, then I got a callback and read with a few people. We did the first five, and then they decided to continue it. They developed families and friends and wives and husbands and kids. I think I've had seven different kids over the course of two-and-a-half years. My eldest I must have had when I was 11, because she's 24.

When you book a commercial don't spend the money until you've got it in your hand—because you have no idea what you are going to end up with.

Patrick Garner is a theatre actor successful in o/c and v/o commercials. He and wife Rosemarie will be relocating to the West Coast later this year.

John Speredakos

I've done my first five years in New York, and I've had some successes. I'm going to turn 30 this summer, which is always a milestone. It feels as if a chapter has ended. My agents would have me believe that I'm hot and that I work all the time. But I know what the reality is. I know that after "Brewster Place" there were six or seven months where I didn't do *anything*. Period. No staged reading. No commercial. No play. No film. No episodic TV. Which was doubly disappointing because you would have thought that after a lead role on a prime-time series you would at least get something. But then last May to this May has no doubt been my best career year. I did a play at The Vineyard, I got the film *Jersey Girl*, I did a great commercial for American Express that Michael Lindsay-Hogg directed, but which, unfortunately, didn't run. Finally the year climaxed with the pilot for "Sirens" getting picked up. All these things combined have given me some interesting perspectives.

I think that the business has changed to the point that in the 90s it's so difficult to make a living, that to not do commercials is very impractical. I have a GE spot that I did the summer I graduated from Rutgers. It was a national, and after having run for two years it was picked up and run for another two. Then they lifted my services from the spot and incorporated it into a whole new spot. It's five years now, and it took me one day to shoot. I don't care what your artistic ideals are; there were many times in the first two or three years when a GE residual would be the sole source of income available to pay my rent. When you can make money that easily and have it last that long it's foolish not to take advantage of that.

I certainly have artistic goals as lofty as anybody's, and as my career changes my relationship with commercials will change. It already has.

I have never waited a table in my life. My first year I worked as a security guard in a hospital on the midnight shift, but beyond that I've had very few survival jobs. I find that people get trapped in their survival jobs so easily. I think it was John Goodman who said, "You have to force yourself to live within your means as quickly as possible. As soon as you can, start making your living as a professional actor even if it's not much of a living." As soon as you can, support yourself entirely as an actor because that's where your self-respect comes from. The problem is that people say, "Well I've got to have a certain amount of money, so I've got to get a survival job." I sort of bit the bullet when I got out of grad school. I moved into a tiny place in East Harlem on 104th Street. I had nothing. I owned nothing. I owed over ten thousand dollars in student loans. I had no money and no furniture. I had no girlfriend and I had no career. The only thing I had was the knowledge that I had nowhere to go but up. That's tremendously liberating if you let it be. I had nothing to rely on but myself, I was in such deep sh— already that I could only go forward. But a lot of actors get trapped in their survival jobs.

I live so cheaply. I generally eat one meal a day at dinner. I work out in my apartment. It's a very self-contained existence, but that has enabled me to survive while other people have priced themselves out of trying to do this.

A series regular on "Sirens" (ABC), John Speredakos starred in the TV show "Brewster Place" (ABC) and had a recurring role on "Kojak." He has appeared in the films School Ties *and* Jersey Girl.

13 □□□ □□□ □□□

Questions and Answers

What Kind of Investment Does Breaking into This Business Require?

For adults, the initial monetary investment will need to cover the cost of a photo session, reproduction of photos and resumes, envelopes and postage, answering machine and/or answering service, telephone bills, wardrobe and maintenance—one or two versatile outfits and cleaning costs—personal grooming—hair, nails, and fitness—trade papers, classes, and transportation. Union dues, bigger wardrobes, and more photo sessions will come later. Combined, these costs can run into several thousands of dollars. If you are hesitant about your potential for success in this business, consult with a reputable commercial instructor, or call up a union franchised talent agency and ask if any of the agents are holding classes or seminars.

Timewise, you must be ready to commit many hours a week for mailings, making the rounds, interviews, classes, and, eventually, auditions and bookings. Auditions could be infrequent—once or twice monthly, or as many as several a day—depending on your type, talent, and level of determination.

For children, the initial investment must cover the cost of getting a good snapshot reproduced and sent out, telephone bills, answering machine or service, and transportation. Before spending any money, find an agent or manager to work with your child to see what his potential is.

How Much Can I Make on the Average Commercial?

The only money you will be guaranteed of making is the session fee, which at scale for on-camera principles is $366. If the commercial runs well or your agent was able to negotiate more money for you, the payback can be much greater. Actors make many thousands of dollars on a single scale commercial that runs well. However, there is no guarantee that once you have shot the commercial it will actually end up on the air or run for any length of time. If it doesn't run, it will either be held, entitling you to a scale payment of $366 per thirteen weeks, or released, in which case you will make no additional money, but be free to work on conflicting spots.

How Much Should I Spend on Pictures?

The average cost for a photo sitting can run anywhere from $150 to $900. Before spending any money, get recommendations for good photographers and interview them. Reproduction costs are usually under $100 for a batch of a hundred or more.

Which Photographers Are Good?

Currently, in New York for adults: Susanna Gold, Jinsey Dauk, Glenn Jussen, Gerard Barnier, Raffi, Joe Henson. In Los Angeles: Joan Lauren, Denise Winters, Robert Strauss. For kids: Jean Poli, Bob Cass, and Susanna Gold, and many others in all categories. A photographer may be hailed as the greatest thing since sliced bread by someone else, and yet not be right for you. All have their own strengths and weaknesses. Do your homework.

Do I Need Commercial Classes?

If you have had no commercial experience or little exposure in front of a camera, yes. If you are rusty or can't seem to book jobs, you may need a brushup or a one-on-one.

Which Commercial Classes Are Good?

In New York: Actors in Advertising, Ruth Lehner, Anita Guererra. In L.A.: Beth Launer and Stuart Robinson, among others. Once again, check around, research.

Should I Put Together a Reel?

Not until you have professional work to your credit.

Should I Join the Union?

Most agents will work with you regardless of whether you are in the union. It is generally a good idea to wait until you have booked your first job to join.

Is It Better to Have an Answering Machine or Answering Service?

It is better to have both. One or the other has been known to fail. Answering machines can malfunction or their owners forget to turn them on, and answering service employees keep people holding, misplace messages, get them wrong, or give them to the wrong people. Having both will lower the margin for error. When you become active, a beeper is a wise investment. If you have an answering machine, make sure you can call into it from an outside phone to check your messages; otherwise it is basically worthless. Whether using a machine or service, you should check your messages frequently, hourly, if possible.

How Do I Know If I Have Star Quality?

Star quality is a very hard thing to define, and even harder to pinpoint. Whether or not you think you have it, work on the other areas that will make you a good commodity and stand you in good stead in the long run, such as talent, technique, attitude, contacts, and appearance.

How Do I Know What My Commercial Type Is?

Research, ask questions, watch TV. Through knowledge and refinement you will eventually be able to perfect your type.

How Do I Become Hot?

Being hot is just as much a state of mind as it is a state of being. No one can make herself hot. The best that you can do is project the appearance of being in demand. If you are confident and upbeat and project excitment and success, people will pick up on your attitude.

Why Won't Anyone Hire Me?

Take a good look at yourself and see whether you are projecting desperation. Conversely, perhaps you are too cocky, arrogant, pushy or complaining. Maybe your energy is too low. Whatever the case, look within before putting the blame elsewhere.

What Is a Clean Contract?

A clean contract means that it must be left intact, without any of the standard provisions crossed out or altered. Many actors don't realize that unless the job is stipulated as being a clean contract, they have the right to cross out the Foreign Use, Theatrical, and Industrial Use clauses found on the back of the job contract. (See the sample commercial employment contract in Appendix B.) If the clauses are not crossed out, the producers are obliged to pay no more than scale for such uses. When they are crossed out, the producers are obliged to come back and negotiate for more money.

What Are Conflicts:

Conflicts, or a commercial's "exclusivity," are those products or services which compete with the sponsor's product or service in the given market area.

What Is the *Mark*?

The mark is a piece of tape on the floor of the audition room, indicating where you should stand so as to be in the camera's line of sight.

How Do I *Slate*?

You merely say your name confidently and succinctly.

How Do I Make Sure I Get Paid on Nonunion Jobs?

Ask to be paid in cash on the set.

How Do I Get my Child into the Business?

Take some good snapshots of your child. Reproduce one that is close up, clearly focused, and transmits a lot of energy. Then mail it to the union franchised agents listed in the *Ross Reports*, or other talent guides, who handle children.

Do I Need a Manager?

If you cannot get an agent, are too busy to handle all the details of your career on your own, need to reposition yourself in the business, are a parent who lives in L.A. or doesn't feel up to handling all the details of your child's career yourself, a manager may be right for you.

Do I Need an Agent to Find Work?

In L.A., yes; otherwise an agent is important, but not essential. Nonunion commercial jobs, commercial print work, and industrials are open to talent with no representation. If you get a callback on a union commercial that you got of your own accord, the casting director will put you through an agent. This is an excellent way of gaining entry into an agent's office.

How Do I Make an Impression on an Agent?

Be knowledgeable about yourself and the business, be confident without being cocky, straightfoward without being pushy. Know who you are and your selling points, present yourself well, and, most of all, be yourself.

Should I Sign or Freelance?

In the beginning you will probably not be given a choice, for no one may want to sign you until you have some momentum going. After you gain a foothold in the business, it is up to you to decide whether working exclusively with one agent or freelancing is in your best interests.

What is the Difference Between First Refusals and General Holds?

Although not formally recognized by the union, technically a first refusal gives the casting director the first right to your time on a given day or period of time. While a *hold* is supposed to equal a booking, first refusals and *general holds* are not quite as formal.

If I Get a Callback or a First Refusal does It Mean that I've Booked the Job and the Rest Is a Mere Formality?

When an agent hears that her client is first choice on a particular job she considers it the kiss of death. With the amount of people involved in making the final decision these days, callbacks, first refusals, or general holds mean nothing more than they would like to see more of you, and that you haven't yet been eliminated from consideration.

How Do I Avoid Nervousness?

Nervousness is primarily a mind game. While there are many ways to combat it, the most effective is the confidence that comes with experience and success.

What Do I Put on Credit Applications Under "Employer?"

If you are a working commercial actor signed to an agency, or if you freelance primarily through one source, you can list your agency or manager as your employer. It is a good idea to alert them ahead of time to the fact that someone may be calling to check your credit.

I'm Told that Commercial Actors are in Danger of Becoming Overexposed. What is it and How Do I Avoid It?

Overexposure occurs when your face appears too frequently in a given area. It is a real fear, and has been known to hurt actors' careers particularly when the area of overexposure is New York or Los Angeles. For this reason, an agent should seek to limit your exposure in these two markets, and warn actors that they feel are headed in that direction. The best safeguard against overexposure is having an agent who sees eye-to-eye with you on the direction and potential of your career, and steers quality and well positioned work your way.

An Agency Wants to Sign Me Across the Board, but They Are Not That Strong Commercially. I Can't Seem to Get Legit Representation Any Other Way. What Are My Options?

The prevailing rule of thumb is that, if an agent wants to work with you for legit, they will work with you regardless of where you place your commercial business. Unfortunately, the reality is that many otherwise good agencies will strongarm actors who are thriving commercially, but in need of legit backing, into signing with them across the board. Whereas this practice is not necessarily fair to the actor, it is how some agencies build up their commercial departments. It is up to you to prioritize your career. If you want to pull back commercially and concentrate on legit, then this situation might work out for you.

How Do I Keep My Picture From Ending Up in the Dead File?

Make sure your picture is of good technical quality, presents you as a clear type, and posseses life and vitality. Its energy should *speak* to an agent. Be enticing in your cover letter, but don't be pushy or state the obvious. Make sure it is short and clearly written or typed. Your resume should be attached to the headshot, not loose, and present you honestly. Make sure your phone number is on it. Send your picture to the proper party, and address that person by name, not as "Dear Sir or Madam," or "To Whom It May Concern."

As an Actor Who Needs to Make a Living, What Other Areas Are Open to Me Besides Commercials?

Industrials and commercial print are two other areas that can greatly supplement an actor's income.

My Agent Keeps Sending Me Out One Particular Way Although I Know I Can Do More. Should I Change Agents?

Before doing anything rash, sit down and have a heart-to-heart talk with your agent. Perhaps you have to prove your abilities before being sent out in those other areas. Get yourself into a show that showcases your talents the way in which you would like to be presented, then get your agent to attend with a casting director. Maybe you aren't being honest with yourself about how you are salable. Get opinions of other people in the industry besides your agent. Maybe casting directors only see you a certain way, as a certain age or a certain type. If they are wrong, you and your agent need to work on getting those opinions changed.

How Do I Get Work On my Own, Without an Agent?

Network, network, network. It can be done, with a lot of perseverance and knowhow. Get to know producers, directors and casting directors. See and be seen.

What Are the Biggest Mistakes Actors Make?

The biggest mistakes are probably not being honest with oneself about one's potential and place within the business, not setting goals, not having a clear idea of where one is going, not being objective about one's appearance or age, not working on craft and technique but sitting back and letting others do the work, not being ready when opportunity strikes, and antagonizing or putting people off with negative or inappropriate attitudes.

What Do I Do When All Else Fails?

You must step back and make an honest assessment of yourself, your goals, your priorities, and your methods.

Are you aiming too high? Do you want fame and fortune and nothing else will do, or is making a living as an actor sufficient?

Take an objective view of yourself. Ask ten people—working actors, casting directors, and agents—what they honestly think of you and your talent.

You may not be packaging yourself appropriately: Your pictures may not be working, your hairstyle may be wrong for you, you may be dressing too young or too old, or your wardrobe may need to be updated. Sometimes a small weight loss or change of hair color can work wonders. Maybe you need a drastic change such as cutting off all your hair and going for a completely new look.

You may need to focus on your method of getting in to meet people. Try a new tactic. Knock on doors if you've only been doing mailings. Try an outlandish snapshot picture of yourself as a postcard. Send an agent a singing telegram and deliver it yourself. Take a full page ad out in the trades. Get your friends to introduce you to their agents.

If you audition but can't seem to get callbacks, then there may be something that you are or are not doing on the audition. Ask the casting directors where you are falling short. Take a one-on-one session with a commercial instructor who also casts. Perhaps you need to back away from the business and do something else for awhile, returning with a new outlook and vigor, or realize that it's not the business for you.

Maybe you need a technique refresher course and to go back to the basics. If you've never taken a commercial class, try one now. If you have had no formal acting training, enroll with a respected instructor.

Whatever the case, try to be objective, don't be afraid to ask questions, and don't be afraid to shake things up a bit. Remember, there are no rules.

If you are sure that this is the business for you, you have tried everything and nothing seems to click, then perhaps you just need to grow into yourself. Sometimes when an actor reaches a certain age, or attains a certain outlook and air about himself, everything falls into place—for no explicable reason. All he can do until he reaches that point is work hard, study and persevere. When it does happen, he'll be ready.

□ □ □
□ □ □
□ □ □

Recommended Reading

Books

Benedict, Larry and Susan Benedict, *The Video Demo Tape: How To Save Money Making A Tape That Gets You Work*. Focal Press, 1992

Bernard, Ian. *Film and Television Acting*. Focal Press, 1993

Blum, Richard A. *Working Actors: The Craft of Television, Film and Stage Performance*. Focal Press, 1989

Bruder, Melissa, Cohn, Lee Michael, Olnek, Madeleine, Pollack, Nathaniel, Previto, Robert, and Zigler, Scott, *A Practical Handbook for the Actor*. Vintage Books, Random House, New York, 1986

Gold, Aggie, *Fresh Faces: Getting Your Child Into Commercials, Television and the Movies* Career Press, 1990

Henry, Mari Lynn and Rogers, Lynn. *How to be a Working Actor* M. Evans and Co., 1989

Padol, Brian A. and Simon, Allen, *The Young Performer's Guide: How to Break Into Show Business*. Betterway Publications, 1990

Sandler, Bernard, and Posner, Steve, *In Front of the Camera: How To Make It and Survive in Movies and Television*. Elsevier-Dutton Publishing, 1981

Searle, Judith, *Getting the Part* Simon & Schuster, 1991

Shurtleff, Michael, *Audition!* Walker & Co. 1978

Small, Edgar, *Agent to Actor*. Samuel French, 1991

White, Hooper, *How To Produce Effective TV Commercials*. NTC Business Books, 1989

Publications

Backstage, New York, published weekly

Dramalogue, Los Angeles, published weekly

The Hollywood Reporter, Los Angeles, published daily

The Madison Avenue Handbook, Peter Glenn Publications, published annually

The Ross Reports, New York, published monthly

The Ross Reports USA, published quarterly

Variety, Los Angeles, published weekly

Working Actors Guide, Los Angeles, published annually

A

Union Offices Nationwide

SCREEN ACTORS GUILD (SAG)

ARIZONA
1616 E. Indian School Road
Pheonix, AZ 85016
(602)265-2712

ATLANTA
1627 Peachtree St., N.E. #210
Atlanta, GA 30309
(404)897-1335

BOSTON
11 Beacon St. Rm 512
Boston, MA 02108
(617)742-2688

CHICAGO
75 East Wacker Dr., 14th Floor
Chicago, IL 60601
(312)372-8081

DALLAS
6060 N. Central Expressway
Suite 302, LB 604
Dallas, TX 75206
(214)363-8300

DENVER
950 S. Cherry St., #502
Denver, CO 80222
(303)757-6226
(800)527-7517

DETROIT
28690 Southfield Rd., #290 A&B
Lathrup Village, MI 48076
(313)559-9540

FLORIDA
2299 Douglas Rd., #200
Miami, FL 33145
(305)444-7677
3393 W. Vine St., Suite 302
Kissimmee, FL 34741
(407)847-4445

HAWAII
949 Kapiolani Blvd.
Honolulu, HI 96814
(808)538-6122

LOS ANGELES
7065 Hollywood Blvd.
Hollywood, CA 90028-6065
(213)856-6612

HOUSTON
2650 Fountainview, #326
Houston, TX 77057
(713)972-1806

NASHVILLE
P.O. Box 021087
Nashville, TN 37212
(615)327-2958

NEVADA
Served through the Denver office

NEW MEXICO
Served through the Denver office

NEW YORK
1515 Broadway, 44th floor
New York, NY 10036
(212)827-1474

PHILADELPHIA
230 S. Broad St., 10th Fl.
Philadelphia, PA 19102
(212) 545-3150

SAN DIEGO
7827 Convoy Ct., #400
San Diego, CA 92111
(619) 278-7695

SAN FRANCISCO
235 Pine St., 11th Fl.
San Francisco, CA 94104
(415) 391-7510

WASHINGTON D.C./BALTIMORE
The Highland House
5480 Wisconsin Ave., #201
Chevy Chase, MD 20815
(301) 657-2560

UTAH
This is an organizing area

*AMERICAN FEDERATION OF
TELEVISION AND RADIO ARTISTS
(AFTRA)*

MAIN OFFICE (NEW YORK)
260 Madison Avenue
New York, NY 10016

LOS ANGELES
6922 Hollywood Boulevard
Los Angeles, CA 90028

CHICAGO
307 North Michigan Avenue
Chicago, IL 60601

LOCAL OFFICES:

ALBANY
341 Northern Boulevard
Albany, NY 12204

ATLANTA
1627 Peachtree Street, NE, #210
Atlanta, GA 30309

BOSTON
11 Beacon Street #512
Boston, MA 02108

BUFFALO
2077 Elmwood Avenue
Buffalo, NY 14207

CLEVELAND
1367 East Sixth Street #229
Cleveland, OH 44114

DALLAS/FORT WORTH
6309 North O'Connor Road #111,
  LB25
Two Dallas Communications
  Complex
Irving, TX 75039-3510

DENVER
950 South Cherry Street #502
Denver, CO 80222

DETROIT
28690 Southfield Road
Lathrup Village, MI 48076

FRESNO
P.O. Box 11961
Fresno, CA 93776

HAWAII
95-314 Kaloapau #18
Mililani, HI 96789

HOUSTON
2620 Fountainview
Houston, TX 77057

KANSAS CITY
406 West Thirty-Fourth Street
Kansas City, MO 46111

MIAMI
20401 NW 2nd Avenue #102
North Miami Beach, FL 33169

MINNEAPOLIS/ST. PAUL
(TWIN CITIES)
15 South 9th Street #400
Minneapolis, MN 55402

NASHVILLE
P.O. Box 121087
1108 17th Avenue South
Nashville, TN 37212

NEW ORLEANS
2475 Canal Street, Suite 108
New Orleans, LA 70119

OMAHA
P.O. Box 31103
Omaha, NB 68131

PEORIA
2907 Springfield Road
East Peoria, IL 61611

PHILADELPHIA
230 South Broad Street—10th Fl.
Philadelphia, PA 10192

PHOENIX
5150 North 16th Street, C-255
Phoenix, AZ 85016

PITTSBURGH
625 Stanwyx Street, The Penthouse
Pittsburgh, PA 15222

PORTLAND
516 SE Morrison, M-3
Portland, OR 97214

RACINE/KENOSHA
929 Fifty-second Street
Kenosha, WI 53140

ROCHESTER
1100 Crossroads Office Building
Rochester, NY 14614

SACRAMENTO/STOCKTON
2413 Capitol Avenue
Sacramento, CA 95816

SAN DIEGO
3045 Rosecrans Street
San Diego, CA 92110

SAN FRANCISCO
100 Bush Street
San Francisco, CA 94104

SCHENECTADY
2040 Hoover Road
Albany, NY 12304

SEATTLE
P.O. Box 9688
601 Valley Street
Seattle, WA 98104

STAMFORD
117 Prospect Street
Stamford, CT 06901

ST. LOUIS
906 Olive Street #1006
St. Louis, MO 63101

TRI STATE CINCINNATI/
COLUMBUS/DAYTON/
INDIANAPOLIS/LOUISVILLE
128 East 6th Street
Cincinnati, OH 45202

WASHINGTON/BALTIMORE
5480 Wisconsin Avenue #201
Chevy Chase, MD 20815

*ACTORS EQUITY ASSOCIATION
(AEA)*

NEW YORK
165 W. 46th St.
New York, NY 10036

LOS ANGELES
6430 Sunset Blvd.
Hollywood CA 90028

CHICAGO
203 N. Wabash Ave.
Chicago, IL 60601

SAN FRANCISCO
100 Bush St.
San Francisco, CA 94104

B

Standard Screen Actors Guild Employment Contract for Television Commercials

ADVERTISING AGENCY _____   PRODUCER _____

COMMERCIAL TITLE(S) AND
CODE NO.(S) _____   PRODUCT _____

Date(s) Worked	Work Time From — To	Meals From — To	Trav. to Loc. From — To	Trav. from Loc. From — To	Fittings, Makeup Test, etc., if on day prior to shooting From — To

Multiple Tracking or Sweetening _____ did occur _____ did not occur

Performer's Signature or Initials: _____ .THIS IS NOT A PART OF THE STANDARD FORM.

- -

EXHIBIT A
STANDARD SCREEN ACTORS GUILD EMPLOYMENT CONTRACT FOR TELEVISION COMMERCIALS

Date _____, 19 _____

Between _____ , Producer, and

_____ , Performer. Producer engages
Performer and Performer agrees to perform services for Producer in television commercials as follows:

Commercial Title(s) and Code No(s) _____ No. of Commercials _____

Check if Applicable
- ☐ Dealer Commercial(s)
  - ☐ Type A
  - ☐ Type B
- ☐ Seasonal Commercial(s)
- ☐ Test or Test Market Commercial(s)
- ☐ Non-Air Commercial(s)
- ☐ Produced for Cable

Such commercial(s) are to be produced by _____ , _____ ,
　　　　　　　　　　　　　　　　　　Advertising Agency　　　　　　　　　　Address

acting as agent for _____ , _____
　　　　　　　　　Advertiser　　　　　　　　　　　Product(s)

City and State in which services rendered: _____   Place of Engagement: _____

( ) Principal Performer	( ) Solo or duo
( ) Stunt Performer	( ) Group-3-5
( ) Specialty Act	( ) Group-6-8
( ) Dancer	( ) Group-9 or more
( ) Singer	( ) Contractor

( ) Signature - solo or duo
( ) Group-Signature-3-5
( ) Group-Signature-6-8
( ) Group-Signature-9 or more
( ) Pilot

Classification: On Camera _____   Off Camera _____   Part to be Played _____

Compensation: _____   Date & Hr. of Engagement: _____

Check if: Flight Insurance ($10) Payable ☐
Wardrobe to be furnished by Producer ☐ by Performer ☐

If furnished by Performer, No. of Costumes @ $15.00 _____ @ $25.00 _____   Total Wardrobe Fee $ _____
　　　　　　　　　　　　　　　(Non-Evening Wear)　　(Evening Wear)

☐ Performer does not consent to the use of his/her services in commercials, made hereunder as dealer commercials payable at dealer commercial rates.

☐ Performer does not consent to the use of his/her services in commercials made hereunder on a simulcast.

The standard provisions printed on the reverse side hereof are a part of this contract. If this contract provides for compensation at SAG minimum, no addition, changes or alterations may be made in this form other than those which are more favorable to the Performer than herein provided. If this contract provides for compensation above SAG minimum, additions may be agreed to between Producer and Performer which do not conflict with the provisions of the SAG Commercials Contract, provided that such additional provisions are separately set forth under "Special Provisions" hereof and signed by the Performer.

Until Performer shall otherwise direct in writing, Performer authorizes Producer to make all payments to which Performer may be entitled hereunder as follows:

☐ To Performer at _____
　　　　　　　　　　　　　(Address)

☐ To Performer c/o _____ at _____
　　　　　　　　　　　　　　　　　　　(Address)

All notices to Performer shall be sent to the address designated above for payments and, if Performer desires, to one other address as follows:

To _____
　　　　　(Name)　　　　　　　　　(Address)

All notices to Producer shall be addressed as follows:

To Producer at _____
　　　　　　　　　　　　(Address)

This contract is subject to all of the terms and conditions of the applicable Commercials Contract. Employer of Record for income tax and unemployment insurance purposes is _____

PRODUCER (NAME OF COMPANY) _____

BY _____   PERFORMER _____

Performer hereby certifies that he/she is 21 years of age or over. (If under 21 years of age this contract must be signed below by a parent or guardian.)

I, the undersigned hereby state that I am the _____ of the above named Performer and do hereby consent and give my permission to this agreement.　　(Mother, Father, Guardian)

_____
(Signature of Parent or Guardian)

SPECIAL PROVISIONS (including adjustments, if any, for Stunt Performers):

Performer acknowledges that he/she has read all the terms and conditions in the Special Provisions section above and hereby agrees thereto.

_____
(Performer)

IMPORTANT PROVISIONS ON BACK. PLEASE READ CAREFULLY.

(W-4 FORM IS ATTACHED HERE)

STANDARD PROVISIONS

1. RIGHT TO CONTRACT

Performer states that to the best of his/her knowledge, he/she has not authorized the use of his/her name, likeness or identifiable voice in any commercial advertising any competitive product or service during the term of permissible use of commericial(s) hereunder and that he/she is free to enter into this contract and to grant the rights and uses herein set forth.

2. EXCLUSIVITY

Performer states that since accepting employment in the commercial(s) covered by this contract, he/she has not accepted employment in nor authorized the use of his/her name or likeness or identifiable voice in any commercial(s) advertising any competitive product or service and that he/she will not hereafter, during the term of permissible use of the commercial(s) for which he/she is employed hereunder, accept employment in or authorize the use of his/her name or likeness or identifiable voice in any commercial(s) advertising any competitive product or service. Unless otherwise bargained for, this paragraph shall not apply to off-camera solo or duo singers or group performers other than name groups or to performers employed in Seasonal Commercials under Section 39 of the SAG Commercials Contract.

3. OTHER USES (Strike "a" or "b" or both if such rights not granted by Performer)

(a)   Foreign Use

Producer shall have the right to the foreign use of the commercial(s) produced hereunder, for which Producer agrees to pay performer not less than the additional compensation provided for in the SAG Commercials Contract. Producer agrees to notify SAG in writing promptly of any such foreign use.

(b)   Theatrical & Industrial Use

Producer shall have the right to the commercial(s) produced hereunder for theatrical and industrial use as defined and for the period permitted in the SAG Commercials Contract, for which Producer shall pay performer not less than the additional compensation therein provided.

4. ARBITRATION

All disputes and controversies of every kind and nature arising out of or in connection with this contract shall be subject to arbitration as provided in Section 56 of the SAG Commercials Contract.

5. PRODUCER'S RIGHTS

Performer acknowledges that performer has no right, title or interest of any kind or nature whatsoever in or to the commercial(s). A role owned or created by Producer belongs to Producer and not to the performer.

C

AFTRA Standard
Employment Contract for
Non-Broadcast Industrials

STANDARD EMPLOYMENT CONTRACT
AFTRA NATIONAL CODE OF FAIR PRACTICE FOR
NON-BROADCAST/INDUSTRIAL/EDUCATIONAL/RECORDED MATERIALS

This Agreement made this _____ day of _____, 19___,
between _____ hereinafter called Producer and
_____ hereinafter called Performer.

1. PROGRAM-Producer engages Performer and Performer agrees to perform
services in a program tentatively entitled _____
_____ to be produced on behalf of _____
                                                            (client)

2. CATEGORY-The initial, primary use of the program shall be (check one):
____ Category 1 (Industrial/Educational)   ____ Category 11 (Point of Purchase)

3. ROLE-Performer shall portray the role of _____

4. TERM-Performer's employment shall be for the continuous period commencing
_____, 19___ and continuing until completion of photography
and recording of said role. EXCEPTION-(for Day Players only): Performer may
be dismissed and recalled once without payment for intervening period providing such period exceeds 5 calendar days and Performer is informed of firm
recall date at time of engagement. If applicable to this contract,
Performer's firm recall date is _____, 19___.

5. COMPENSATION-Producer employs Performer as: ___On-Camera ___ Off-Camera
                ___On-Camera Narrator/Spokesperson
( ) Day Performer   ( ) Singer, Solo/Duo ( ) General Extra Performer
( ) 3-Day Performer   ( ) Singer, Group   ( ) Special Ability Extra Performer
( ) Weekly Performer ( ) Singer, Step Out ( ) Silent Bit Extra Performer
                                            ( ) Non-Principal Performer
at the salary of $_____ per _____.
Off-Camera Performers only, for each additional 1/2 hour _____.
Payment shall be sent to the appropriate AFTRA office in the city nearest the
recording site:_____.

6. ADDITIONAL COMPENSATION FOR SUPPLEMENTAL USE-Producer may acquire the
following supplemental use rights by the payment of the indicated amounts
(Check appropriate item(s) below.)

	At time of employment or within 90 days of completion of principal photography	Beyond 90 Days
	(% of total applicable salary)	(% of total app. salary)
A. Cable TV		
Basic Cable (3 years)	15%	65%
B. Non-Network TV Unlimited Runs	75%	125%
C. Theatrical Exhibition,		
Unlimited Runs	100%	150%
D. Foreign TV Unlimited		
Runs outside U.S. & Canada	25%	75%
E. Integration and/or		
Customization	100%	150%
F. Sales and/or Rental		
To Industry	15%	25%
G. "Package" rights to		
A,B,C,D,E,& F above	200%	Not Available

7. WARDROBE-(check one)-Wardrobe will be furnished by:
        [] Producer   [] Performer
If performer furnishes own wardrobe, the following fees shall apply for each
two-day period or portion thereof: Ordinary Wardrobe $_____ ($15.00 min.);
Evening or Formal Wardrobe $_____ ($25.00 min.)

8. SPECIAL PROVISIONS-

9. GENERAL- All terms and conditions of the AFTRA National Code of Fair
Practice for Non-Broadcast/Industrial/Educational/Recorded Materials shall be
applicable to such employment.

Producer _____   Performer _____

by _____   Soc. Sec. _____
              Name
_____   Address _____
              Title

D

National Conference of Personal Managers Code of Ethics

A personal manager is one who has experience and knowledge of the many facets of the entertainment field. Said manager agrees to use this specialized knowledge to guide, advance and promote the careers of clients who retain the manager's professional services. A manager is to find, develop new talent and create opportunities for the artists they represent. A personal manager is a liaison between the artists he/she represents and booking agents, the entertainment industry and the general public.

For the privilege of a personal manager to be a member of the National Conference of Personal Managers, he/she must . . .

1. Have as his/her primary occupation the management of entertainers, performing artists and creative personalities.
2. Always deal honestly and fairly with his/her clients.
3. Not derive personal gains at the expense of his/her clients.
4. Treat relationships with clients in a confidential manner.
5. Not encourage or induce an artist to breach an existing personal management contract; not represent an artist while the artist is under a valid contract to another member in good standing of the National Conference of Personal Managers, except by written agreement with all parties concerned. In the event an agreement cannot be reached, before any other action is taken, the matter must be referred to the NCOPM arbitration committee.
6. Be proud of the management profession and the National Conference of Personal Managers. NCOPM members shall not publicly or privately disparage a fellow member of the National Conference of Personal Managers. In the event there is a difference of opinion between

any members, the matter shall be referred to the appropriate committee for arbitration.

7. To the extent that it does not conflict with the best interest of their clients, members of the Conference should exchange information wherever possible and advisable.